Cambridge Lower Secondary

Science

STAGE 8: STUDENT'S BOOK

Gemma Young, Mark Levesley,
Beverly Rickwood, Lucy Hawkins, Stuart Lloyd

Collins

William Collins' dream of knowledge for all began with the publication of his first book in 1819. A self-educated mill worker, he not only enriched millions of lives, but also founded a flourishing publishing house. Today, staying true to this spirit, Collins books are packed with inspiration, innovation and practical expertise. They place you at the centre of a world of possibility and give you exactly what you need to explore it.

Collins. Freedom to teach.

Published by Collins
An imprint of HarperCollins*Publishers*
The News Building
1 London Bridge Street
London
SE1 9GF

Browse the complete Collins catalogue at
www.collins.co.uk

10 9 8 7 6 5 4

ISBN 978-0-00-825466-7

MIX
Paper from responsible sources
FSC™ C007454

This book is produced from independently certified FSC paper to ensure responsible forest management.

For more information visit:
www.harpercollins.co.uk/green

British Library Cataloguing in Publication Data
A catalogue record for this publication is available from the British Library.

Authors: Gemma Young, Mark Levesley, Beverly Rickwood, Lucy Hawkins, Stuart Lloyd
Development editors: Jane Glendening, Gillian Lindsey, Gina Walker
Team leaders: Mark Levesley, Peter Robinson, Aidan Gill
Commissioning project manager: Susan Lyons
Commissioning editors: Joanna Ramsay, Rachael Harrison
In-house editor: Natasha Paul
Copyeditor: Rebecca Ramsden
Proofreader: Mitch Fitton
Technical checker: Sarah Binns
Indexer: Jouve India Private Limited
Photo researcher: Alison Prior
Illustrator: Jouve India Private Limited
Cover designer: Gordon MacGilp
Cover artwork: Maria Herbert-Liew
Internal designer: Jouve India Private Limited
Typesetter: Jouve India Private Limited
Production controller: Tina Paul
Printed and bound by: Grafica Veneta in Italy

Contents

Chapter 7 • Light

Chapter 8 • Sound

Chapter 9 • Measuring motion

Chapter 10 • Magnets

How to use this book

This book is designed to challenge you to go beyond the content you need to learn on your course. Have a go at the questions in dark green, blue and orange to challenge yourself, and read more about the scientific world in the discovering sections.

Chapter 1 . Topic 1

Food from plants

Learning outcomes

- To describe how plants make their own food, using photosynthesis
- To explain why plants need a continuous source of water
- To explain how leaves are adapted for photosynthesis

Starting point

You should know that...	You should be able to...
Plants make their own food, using chloroplasts	Explain how scientists think up ideas, make predictions, and collect evidence to test their ideas
Plants need light and water to grow	Use tables, bar charts and line graphs
Plants are producers, and make the energy-containing substances needed by all the other organisms in a food chain	

Resources for plants

Plants need these resources from their habitats:

- light
- carbon dioxide gas
- water
- warmth
- small amounts of substances called **mineral salts**.

1 Why did no plants grow under the trampoline in figure 1.1?

2 Why are there not many plants in sandy deserts?

Key term

mineral salt: type of substance in the soil that plants need small amounts of. Often just called a 'mineral'.

1.1 There used to be a trampoline here. Plants did not grow under it.

Photosynthesis

Plants make their own food using a process called **photosynthesis**. This process uses chemical substances (**reactants** or **raw materials**) to make new chemical substances (**products**). A process in which reactants form products is a **chemical reaction**.

In photosynthesis, carbon dioxide gas and water combine to form **glucose** and oxygen.

3 a) What are the raw materials for photosynthesis?

b) What are the products of photosynthesis?

1.2 There are not many plants in the Rub al Khali desert (which is in Oman, Saudi Arabia, United Arab Emirates and Yemen).

Food from plants 3

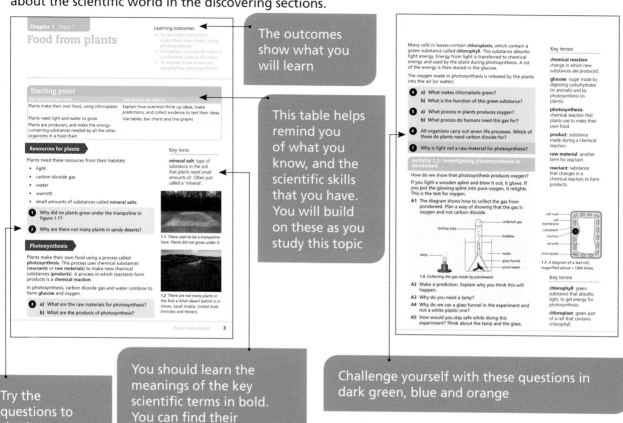

The outcomes show what you will learn

This table helps remind you of what you know, and the scientific skills that you have. You will build on these as you study this topic

Many cells in leaves contain **chloroplasts**, which contain a green substance called **chlorophyll**. This substance absorbs light energy. Energy from light is transferred to chemical energy and used by the plant during photosynthesis. A lot of the energy is then stored in the glucose.

The oxygen made in photosynthesis is released by the plants into the air (or water).

4 a) What makes chloroplasts green?

b) What is the function of this green substance?

5 a) What process in plants produces oxygen?

b) What process do humans need this gas for?

6 All organisms carry out seven life processes. Which of these do plants need carbon dioxide for?

7 Why is light *not* a raw material for photosynthesis?

Activity 1.1: Investigating photosynthesis in pondweed

How do we show that photosynthesis produces oxygen?

If you light a wooden splint and blow it out, it glows. If you put the glowing splint into pure oxygen, it relights. This is the test for oxygen.

A1 The diagram shows how to collect the gas from pondweed. Plan a way of showing that the gas is oxygen and not carbon dioxide.

boiling tube, collected gas, bubbles, lamp, water, glass funnel, pond-weed

1.4 Collecting the gas made by pondweed.

A2 Make a prediction. Explain why you think this will happen.

A3 Why do you need a lamp?

A4 Why do we use a glass funnel in the experiment and not a white plastic one?

A5 How would you stay safe while doing this experiment? Think about the lamp and the glass.

Key terms

chemical reaction: change in which new substances are produced.

glucose: sugar made by digesting carbohydrates (in animals) and by photosynthesis (in plants).

photosynthesis: chemical reaction that plants use to make their own food.

product: substance made during a chemical reaction.

raw material: another term for reactant.

reactant: substance that changes in a chemical reaction to form products.

cell wall, cell membrane, cytoplasm, nucleus, vacuole, chloroplast

1.3 A diagram of a leaf cell, magnified about × 1000 times.

Key terms

chlorophyll: green substance that absorbs light, to get energy for photosynthesis.

chloroplast: green part of a cell that contains chlorophyll.

Try the questions to check your understanding

You should learn the meanings of the key scientific terms in bold. You can find their meanings in the margin and in the glossary (near the end of the book)

Challenge yourself with these questions in dark green, blue and orange

Discover more about where scientific ideas have come from and how they are used around the world now

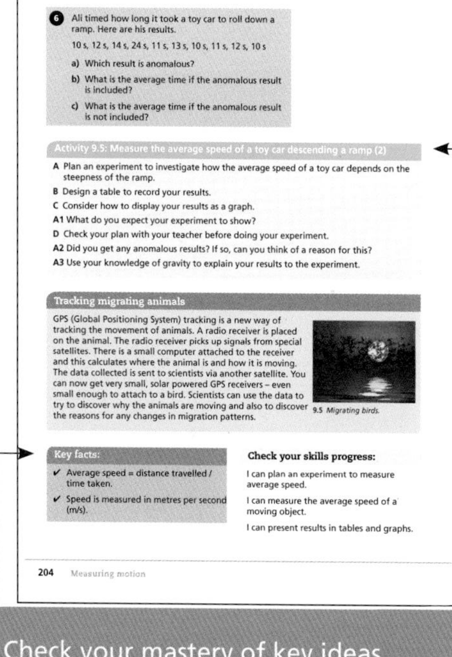

6 Ali timed how long it took a toy car to roll down a ramp. Here are his results.

10 s, 12 s, 14 s, 24 s, 11 s, 13 s, 10 s, 11 s, 12 s, 10 s

a) Which result is anomalous?

b) What is the average time if the anomalous result is included?

c) What is the average time if the anomalous result is not included?

Activity 9.5: Measure the average speed of a toy car descending a ramp (2)

A Plan an experiment to investigate how the average speed of a toy car depends on the steepness of the ramp.

B Design a table to record your results.

C Consider how to display your results as a graph.

A1 What do you expect your experiment to show?

D Check your plan with your teacher before doing your experiment.

A2 Did you get any anomalous results? If so, can you think of a reason for this?

A3 Use your knowledge of gravity to explain your results to the experiment.

Tracking migrating animals

GPS (Global Positioning System) tracking is a new way of tracking the movement of animals. A radio receiver is placed on the animal. The radio receiver picks up signals from special satellites. There is a small computer attached to the receiver and this calculates where the animal is and how it is moving. The data collected is sent to scientists via another satellite. You can now get very small, solar powered GPS receivers – even small enough to attach to a bird. Scientists can use the data to try to discover why the animals are moving and also to discover the reasons for any changes in migration patterns.

9.5 Migrating birds.

Key facts:

- ✔ Average speed = distance travelled / time taken.
- ✔ Speed is measured in metres per second (m/s).

Check your skills progress:

I can plan an experiment to measure average speed.

I can measure the average speed of a moving object.

I can present results in tables and graphs.

204 Measuring motion

Try out the science for yourself with step-by-step activities

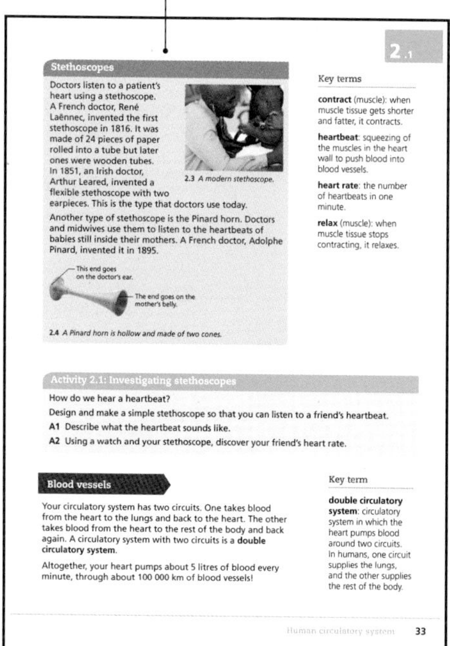

2.1

Stethoscopes

Doctors listen to a patient's heart using a stethoscope. A French doctor, René Laënnec, invented the first stethoscope in 1816. It was made of 24 pieces of paper rolled into a tube but later ones were wooden tubes. In 1851, an Irish doctor, Arthur Leared, invented a flexible stethoscope with two earpieces. This is the type that doctors use today.

Another type of stethoscope is the Pinard horn. Doctors and midwives use them to listen to the heartbeats of babies still inside their mothers. A French doctor, Adolphe Pinard, invented it in 1895.

2.3 A modern stethoscope.

This end goes on the doctor's ear.
The end goes on the mother's belly.

2.4 A Pinard horn is hollow and made of two cones.

Key terms

contract (muscle): when muscle tissue gets shorter and fatter, it contracts.

heartbeat: squeezing of the muscles in the heart wall to push blood into blood vessels.

heart rate: the number of heartbeats in one minute.

relax (muscle): when muscle tissue stops contracting, it relaxes.

Activity 2.1: Investigating stethoscopes

How do we hear a heartbeat?

Design and make a simple stethoscope so that you can listen to a friend's heartbeat.

A1 Describe what the heartbeat sounds like.

A2 Using a watch and your stethoscope, discover your friend's heart rate.

Blood vessels

Your circulatory system has two circuits. One takes blood from the heart to the lungs and back to the heart. The other takes blood from the heart to the rest of the body and back again. A circulatory system with two circuits is a **double circulatory system**.

Altogether, your heart pumps about 5 litres of blood every minute, through about 100 000 km of blood vessels!

Key term

double circulatory system: circulatory system in which the heart pumps blood around two circuits. In humans, one circuit supplies the lungs, and the other supplies the rest of the body.

Human circulatory system 33

Check your mastery of key ideas and skills with this list

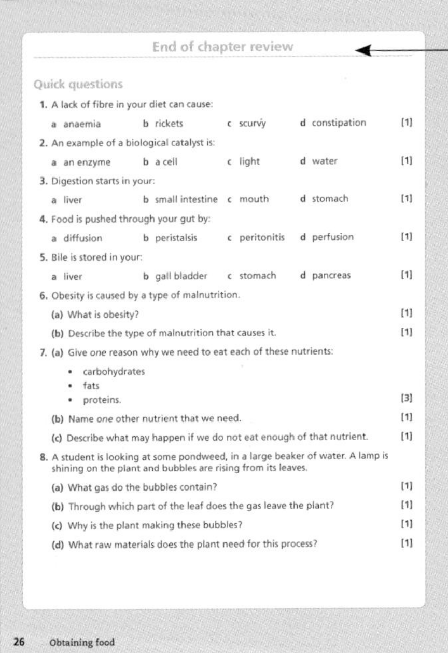

End of chapter review

Quick questions

1. A lack of fibre in your diet can cause:
   a anaemia b rickets c scurvy d constipation [1]
2. An example of a biological catalyst is:
   a an enzyme b a cell c light d water [1]
3. Digestion starts in your:
   a liver b small intestine c mouth d stomach [1]
4. Food is pushed through your gut by:
   a diffusion b peristalsis c peritonitis d perfusion [1]
5. Bile is stored in your:
   a liver b gall bladder c stomach d pancreas [1]
6. Obesity is caused by a type of malnutrition.
   (a) What is obesity? [1]
   (b) Describe the type of malnutrition that causes it. [1]
7. (a) Give *one* reason why we need to eat each of these nutrients:
   - carbohydrates
   - fats
   - proteins. [3]

   (b) Name *one* other nutrient that we need. [1]
   (c) Describe what may happen if we do not eat enough of that nutrient. [1]
8. A student is looking at some pondweed, in a large beaker of water. A lamp is shining on the plant and bubbles are rising from its leaves.
   (a) What gas do the bubbles contain? [1]
   (b) Through which part of the leaf does the gas leave the plant? [1]
   (c) Why is the plant making these bubbles? [1]
   (d) What raw materials does the plant need for this process? [1]

26 Obtaining food

This section helps you check that you understand the ideas and can apply them to new situations

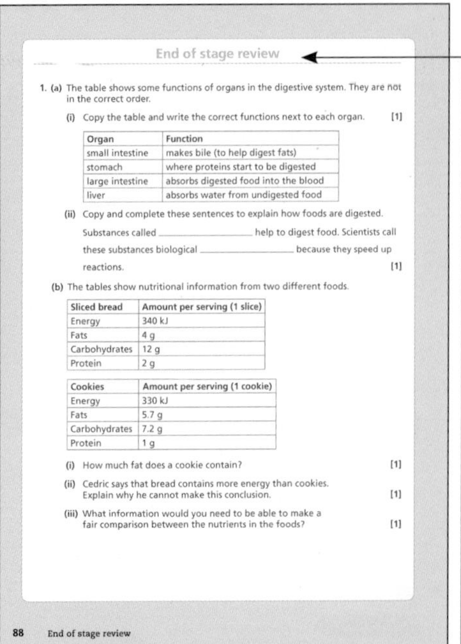

End of stage review

1. (a) The table shows some functions of organs in the digestive system. They are not in the correct order.
   (i) Copy the table and write the correct functions next to each organ. [1]

Organ	Function
small intestine	makes bile (to help digest fats)
stomach	where proteins start to be digested
large intestine	absorbs digested food into the blood
liver	absorbs water from undigested food

   (ii) Copy and complete these sentences to explain how foods are digested.
   Substances called ________ help to digest food. Scientists call these substances biological ________ because they speed up reactions. [1]

   (b) The tables show nutritional information from two different foods.

Sliced bread	Amount per serving (1 slice)
Energy	340 kJ
Fats	4 g
Carbohydrates	12 g
Protein	2 g

Cookies	Amount per serving (1 cookie)
Energy	330 kJ
Fats	5.7 g
Carbohydrates	7.2 g
Protein	1 g

   (i) How much fat does a cookie contain? [1]
   (ii) Cedric says that bread contains more energy than cookies. Explain why he cannot make this conclusion. [1]
   (iii) What information would you need to be able to make a fair comparison between the nutrients in the foods? [1]

88 End of stage review

At the end of the stage, try the end of stage review! This contains questions on all the chapters

Glossary

Biology

absorb: to take in or soak up.

addictive: substance that makes people feel that they must have it.

adolescence: the life stage in humans that usually happens between the ages of 11 and 18. During this stage, people go through many emotional and physical changes.

aerobic respiration: respiration that requires oxygen to release energy from glucose.

alimentary canal: tube that runs from your mouth to your anus.

alveolus: tiny, pocket-shaped structure in lungs where gaseous exchange happens. The plural is alveoli.

amniotic fluid: liquid that surrounds the foetus in the uterus.

anus: last organ in the alimentary canal. Faeces leave your body here.

aorta: large artery that leaves the left ventricle of your heart.

artery: thick-walled blood vessel that carries blood away from the heart.

atria: chambers at the top of your heart. You have a left atrium and a right atrium.

balanced diet: eating many different foods to get the correct amounts of nutrients.

bile: liquid that helps fat-digesting enzymes to work.

biomass: mass of all the compounds in an organism, which it has made.

blood: liquid organ that carries substances around the body.

blood vessels: tube-shaped organs that carry blood around the body.

breathing: movements of muscles in your respiratory system that cause air to move in and out of your lungs.

breathing rate: the number of times you inhale and exhale in one minute.

bronchioles: small tubes leading from the bronchus in a lung.

bronchus: large tube leading from the trachea into a lung. Plural is bronchi.

cancer: when cells in a tissue start to make many copies of themselves very quickly.

capillary: tiny blood vessel that carries blood from arteries to veins.

carbohydrate: compound made from carbon, hydrogen and oxygen.

catalyst: substance that speeds up a chemical reaction.

cervix: the neck of the uterus.

chamber: space inside the heart that fills with blood and empties again.

chemical digestion: digestion that is done by chemical substances.

chemical reaction: change in which new substances are produced.

chest: area inside the body between the ribcage, neck, backbone and diaphragm.

chlorophyll: green substance that absorbs light, to get energy for photosynthesis.

chloroplast: green part of a plant cell that contains chlorophyll.

cilia: waving strands that stick out of some cells.

ciliated epithelial cell: specialised cell with waving cilia to sweep mucus along.

circulation: movement of blood around the body.

circulatory system: group of organs that gets blood around the body.

clot: thick mass of blood cells, stuck together.

compound: substance made from elements.

constipation: when your intestines become blocked.

contract (muscle): when muscle tissue gets shorter and fatter, it contracts.

Glossary 231

You can look up definitions for key terms in the glossary

Biology

1

Chapter 1

Obtaining food

What's it all about?

All organisms need nutrition. Plants make their own food (glucose). They use this to make all the other substances they need. To make some of these substances, plants need small amounts of mineral salts from the soil. Venus flytraps live in soils that are low in mineral salts. So, the Venus flytrap catches insects. It slowly digests them, using substances called enzymes. In this way, it gets mineral salts.

You will learn about:

- The raw materials that plants need for photosynthesis
- The products of photosynthesis
- Balanced diets and why we need different types of nutrients
- What happens when we don't get enough of a nutrient
- How diet and fitness affect our bodies
- The organs in the digestive system and how the digestive system works
- The function of enzymes

You will build your skills in:

- Planning investigations to test ideas
- Making and explaining predictions using scientific knowledge and understanding
- Identifying hazards and planning to control risks
- Identifying variables and planning to control some variables
- Using equipment correctly and taking accurate measurements
- Making calculations
- Interpreting data from secondary sources
- Presenting results using tables and charts

Chapter 1 . Topic 1

Food from plants

Learning outcomes

- To describe how plants make their own food, using photosynthesis
- To explain why plants need a continuous source of water
- To explain how leaves are adapted for photosynthesis

Starting point

You should know that...	You should be able to...
Plants make their own food, using chloroplasts	Explain how scientists think up ideas, make predictions, and collect evidence to test their ideas
Plants need light and water to grow	Use tables, bar charts and line graphs
Plants are producers, and make the energy-containing substances needed by all the other organisms in a food chain	

Resources for plants

Plants need these resources from their habitats:

- light
- carbon dioxide gas
- water
- warmth
- small amounts of substances called **mineral salts**.

1 Why did no plants grow under the trampoline in figure 1.1?

2 Why are there not many plants in sandy deserts?

Key term

mineral salt: type of substance in the soil that plants need small amounts of. Often just called a 'mineral'.

1.1 *There used to be a trampoline here. Plants did not grow under it.*

1.2 *There are not many plants in the Rub al Khali desert (which is in Oman, Saudi Arabia, United Arab Emirates and Yemen).*

Photosynthesis

Plants make their own food using a process called **photosynthesis**. This process uses chemical substances (**reactants** or **raw materials**) to make new chemical substances (**products**). A process in which reactants form products is a **chemical reaction**.

In photosynthesis, carbon dioxide gas and water combine to form **glucose** and oxygen.

a) What are the raw materials for photosynthesis?

b) What are the products of photosynthesis?

Many cells in leaves contain **chloroplasts,** which contain a green substance called **chlorophyll**. This substance absorbs light energy. Energy from light is transferred to chemical energy and used by the plant during photosynthesis. A lot of the energy is then stored in the glucose.

The oxygen made in photosynthesis is released by the plants into the air (or water).

4 **a)** What makes chloroplasts green?

b) What is the function of this green substance?

5 **a)** What process in plants produces oxygen?

b) What process do humans need this gas for?

6 All organisms carry out seven life processes. Which of these do plants need carbon dioxide for?

7 Why is light *not* a raw material for photosynthesis?

Activity 1.1: Investigating photosynthesis in pondweed

How do we show that photosynthesis produces oxygen?

If you light a wooden splint and blow it out, it glows. If you put the glowing splint into pure oxygen, it relights. This is the test for oxygen.

A1 The diagram shows how to collect the gas from pondweed. Plan a way of showing that the gas is oxygen and not carbon dioxide.

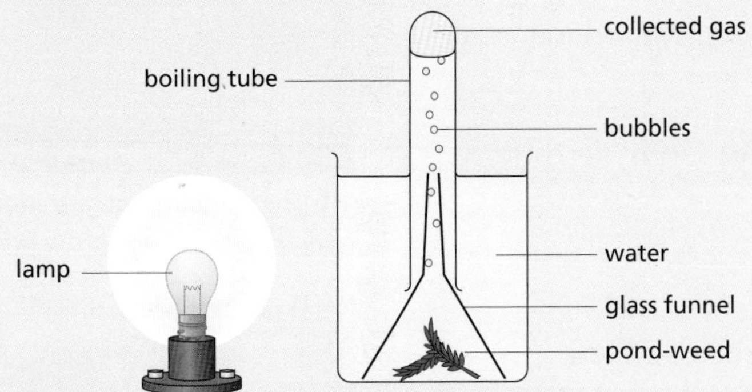

1.4 *Collecting the gas made by pondweed.*

A2 Make a prediction. Explain why you think this will happen.

A3 Why do you need a lamp?

A4 Why do we use a glass funnel in the experiment and not a white plastic one?

A5 How would you stay safe while doing this experiment? Think about the lamp and the glass.

Key terms

chemical reaction: change in which new substances are produced.

glucose: sugar made by digesting carbohydrates (in animals) and by photosynthesis (in plants).

photosynthesis: chemical reaction that plants use to make their own food.

product: substance made during a chemical reaction.

raw material: another term for reactant.

reactant: substance that changes in a chemical reaction to form products.

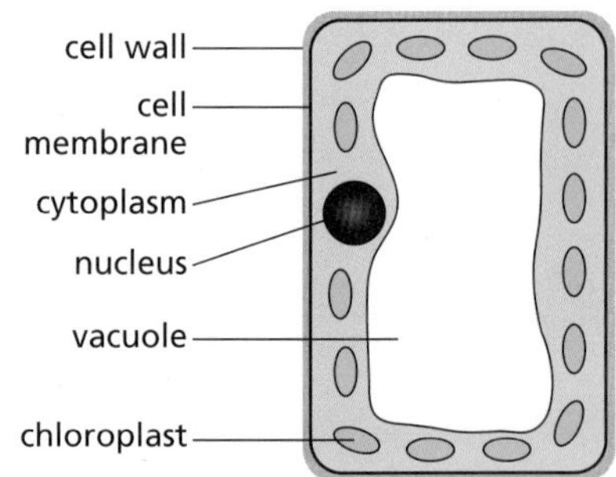

1.3 *A diagram of a leaf cell, magnified about × 1000 times.*

Key terms

chlorophyll: green substance that absorbs light, to get energy for photosynthesis.

chloroplast: green part of a cell that contains chlorophyll.

Artificial photosynthesis

Working in Italy 100 years ago, Giacomo Ciamician had an idea to use artificial photosynthesis to make fuels in countries that did not have oil or coal.

In 2016, American scientists made an 'artificial leaf', to produce oxygen and hydrogen gas using water and the power of sunlight. The hydrogen is mixed with carbon dioxide, which bacteria convert into a liquid fuel. These scientists are now working with others from all over the world, setting up a Global Project on Artificial Photosynthesis. They want to produce cheap fuels that do not cause too much pollution.

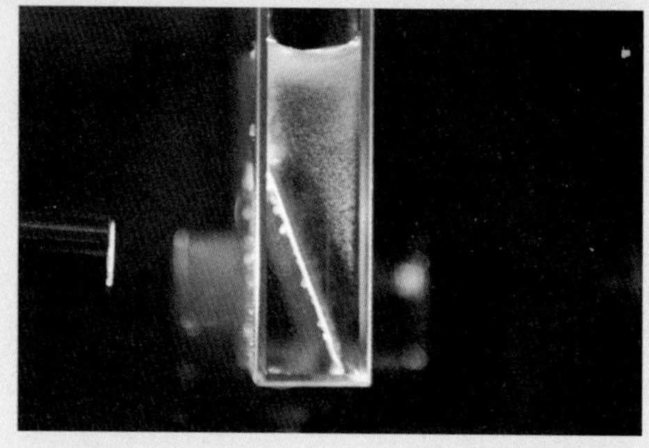

1.5 *This artificial leaf makes bubbles of oxygen and hydrogen gases. Hydrogen is used as a fuel.*

Adaptations of a leaf

Leaves are adapted for photosynthesis by containing different types of cells.

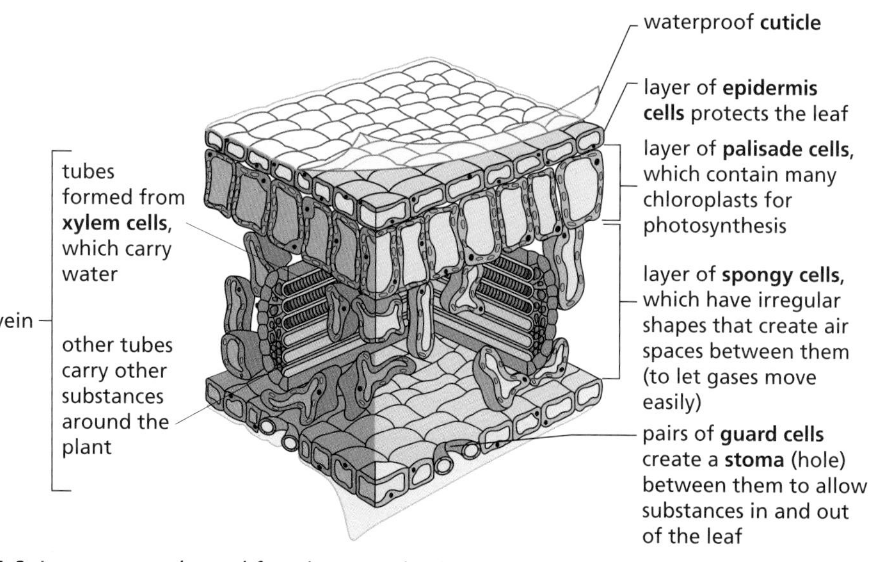
waterproof **cuticle**
layer of **epidermis cells** protects the leaf
layer of **palisade cells**, which contain many chloroplasts for photosynthesis
layer of **spongy cells**, which have irregular shapes that create air spaces between them (to let gases move easily)
pairs of **guard cells** create a **stoma** (hole) between them to allow substances in and out of the leaf
vein
tubes formed from **xylem cells**, which carry water
other tubes carry other substances around the plant

1.6 *Leaves are adapted for photosynthesis.*

8 What is the function of spongy cells in a leaf?

9 **a)** In which type of cell does most photosynthesis occur?

b) How are these cells adapted for photosynthesis?

10 What gas from the air needs to enter leaves through a stoma, for photosynthesis?

Key terms

cuticle: waterproof covering on leaves.

epidermis cell: cell that forms an outer covering of a leaf, to protect the leaf.

guard cell: cell that helps form a stoma in a leaf, to allow gases in and out.

palisade cell: cell that contains many chloroplasts for photosynthesis.

spongy cell: irregularly shaped cell that helps form air spaces in a leaf.

stoma: hole in a leaf, formed between two guard cells. The plural is stomata.

xylem cell: plant cell that is adapted to form hollow tubes to transport water.

11 Which cells bring the liquid raw material for photosynthesis into a leaf?

12 Describe how oxygen gets into the air from inside a leaf.

Plant veins contain tubes that carry substances to and from a leaf. Xylem cells in a vein carry water into a leaf and to its cells. Some of the water evaporates from the cells and becomes a gas. This water vapour escapes from the leaves through small holes. One hole is called a stoma and the plural is **stomata**. Plants need a constant supply of water because they lose water through their stomata.

Key term

plant vein: tube containing smaller tubes that carry substances around a plant.

13 What do we call water when it is a gas?

14 Describe how a palisade cell gets the reactants for photosynthesis.

15 How do plants lose water from their leaves?

16 Pairs of guard cells open and close the stomata between them. In some plants, stomata close if it gets very hot. Explain how this helps the plant.

Biomass

A **compound** is a substance made from simpler substances called **elements**. Carbon, hydrogen and oxygen are all elements.

Carbohydrates are a group of compounds made from carbon, hydrogen and oxygen. Some carbohydrates are small and we call them sugars. Glucose is a **sugar**.

Plants make all their compounds from glucose and mineral salts. A plant's **biomass** is the mass of all these compounds.

17 Carbon dioxide is a compound. What does this mean?

18 **a)** Which elements does glucose contain?

b) By what process does a plant produce glucose?

An important compound that plants make is a large carbohydrate, called **starch**. Plants use starch as a store of energy. Plants use glucose to make starch in chloroplasts, during photosynthesis. At night, when photosynthesis stops, the plant uses the starch to make other compounds, which it transports out of the leaves.

Key terms

biomass: mass of all the compounds in an organism, which it has made.

carbohydrate: compound made from carbon, hydrogen and oxygen.

compound: substance made from elements.

element: substance that contains only one type of atom; it cannot be split into anything simpler.

starch: large carbohydrate, which plants use to store energy.

sugar: type of small carbohydrate.

19 Why does photosynthesis stop at night?

20 **a)** At which of these times would you find most starch in a leaf?

- At the start of a day
- In the middle of a day
- At the end of a day

b) Explain your choice in part a.

21 A plant's biomass does not include the water in it. Explain why not.

We test for starch using **iodine solution**. When we add iodine solution to starch, it changes from orange to a blue-black colour. Before we test a leaf, we remove the chlorophyll (by heating in a liquid called ethanol). Removing the green colour shows up the colour of the iodine better.

Key term

iodine solution: liquid that turns from orange to blue-black when added to starch.

We boil the leaves in water (to make them soft).

We then put them in hot ethanol (to remove the green colour).

We wash the leaves in water.

We add iodine solution to the leaves.

1.7 *Testing leaves for starch.*

22 Look at figure 1.7.

a) Which leaf on the white tile contains starch?

b) Explain your answer to part a.

23 A scientist puts a plant in a dark place for two days. The scientist then tests the leaves for starch.

a) Describe how the scientist would do the test.

b) Predict what will happen.

c) Explain your prediction.

Hazards and risks

A **hazard** is the harm that something may cause. The chance of a hazard causing harm is a **risk**. When doing experiments, we need to control the risks (reduce the chances of harm). The table shows how to control the risks when using iodine solution and ethanol.

	Hazards	Controlling the risks
Iodine solution	stains skin and clothing	do not touch iodine solution keep clothes away from iodine solution
	stings if it gets in your eyes	wear eye protection
	toxic to the aquatic environment	avoid release to the environment
Ethanol	catches fire very easily	do not use near flames
	irritates your eyes	wear eye protection

24 Describe how to stop iodine solution getting in your eyes.

25 Describe how to control the risks of using ethanol.

26 When a gas burner heats things, it gets very hot.

a) Describe the hazard of using a gas burner.

b) Suggest *one* way of controlling the risk.

Key terms

hazard: harm that something may cause.

risk: chance of a hazard causing harm.

Key term

variable: something that may change in an experiment.

Activity 1.2: Investigating photosynthesis in leaves 1

How do we show that photosynthesis requires light?

A **variable** is anything that is able to change in an experiment. In an experiment, we change some variables, we stop others from changing and we measure others. In this experiment, the variable that we change is the amount of light.

You will cover some parts of leaves on a plant, and put it in a sunny place. After two days, you will test the leaves for starch.

A1 Write a plan for this experiment.

A2 Describe how you would control the risks when doing this experiment.

A3 Make a prediction. Explain why you think this will happen.

A4 What variable are you measuring?

1.8 *Covering part of a leaf with foil.*

Activity 1.3: Investigating photosynthesis in leaves 2

How do we show that photosynthesis requires chlorophyll?

In this experiment, the variable that we change is the amount of chlorophyll.

A 'variegated leaf' has white patches, in which there is no chlorophyll. You will use variegated leaves from a plant that has been in a sunny place for two days. You will test the leaves for starch.

1.9 *A variegated leaf.*

A1 Write a plan for this experiment.

A2 Describe how you would control the risks when doing this experiment.

A3 Make a prediction. Explain why you think this will happen.

A4 What variable are you measuring?

Key facts:

- ✔ Photosynthesis is a chemical reaction in which carbon dioxide and water turn into glucose and oxygen.
- ✔ Chloroplasts contain green chlorophyll, which absorbs light for photosynthesis.
- ✔ A plant uses glucose and mineral salts (from the soil) to make all its compounds.
- ✔ The mass of the compounds that a plant has made for itself is its biomass.
- ✔ Plants make starch to store energy.
- ✔ We test for starch using iodine solution, which turns from orange to blue-black.

Check your skills progress:

I can plan investigations to test ideas.

I can identify hazards.

I can plan to control risks in investigations.

I can identify variables in investigations.

I can make and explain predictions.

A balanced diet

Learning outcomes

- To describe what a balanced diet contains
- To explain why our bodies need carbohydrates, fats, proteins, vitamins, minerals, fibre and water
- To describe how diet and exercise affect our bodies

Starting point

You should know that...	You should be able to...
Animals and humans need nutrition	Explain how scientists think up ideas, make predictions, and collect evidence to test their ideas
Humans eat many different plants and animals	Plan ways to control risks during investigations
	Use tables, bar charts and line graphs

Diets

Your **diet** is what you eat and drink. You need food for:

- energy
- growth and repair
- health.

A **nutrient** is any substance needed for energy or used as a raw material to make other substances. Your diet should contain the following nutrients:

- **proteins** (for growth and repair)
- carbohydrates (for energy)
- **fats** (also called **lipids**, and used to store energy)
- **vitamins** and **minerals** (for health, growth and repair).

Water is not a nutrient but is important for dissolving and carrying substances around your body. You also need it for sweating, to keep you cool.

Fibre is not a nutrient but you need it to keep your intestines (gut) healthy. Fibre forms most of your solid waste or **faeces**. A lack of fibre may cause **constipation**, when your gut becomes blocked. Good sources of fibre include wholemeal bread, brown or wholegrain rice, cereals, lentils, nuts and fruits.

Key terms

diet: what you normally eat and drink.

fats: nutrients needed by your body to store energy.

lipids: another word for fats.

minerals: nutrients that living organisms need in small amounts for health, growth and repair. Also called mineral salts.

nutrient: substance you need in your diet for energy or as a raw material.

proteins: nutrients you need for growth and repair.

vitamins: nutrients you need in small amounts for health, growth and repair.

1. List three nutrients that your body needs.
2. a) Why is fibre not a nutrient?
   b) Why does your body need fibre?
   c) List two good sources of fibre in your diet.
3. Give an example of a carbohydrate.

1.10 *These foods are good sources of fibre.*

Food labels

Food labels show the amounts of nutrients and fibre in foods. They also show the energy in foods. Your body needs energy for growing, moving and keeping warm.

We measure energy in units called **joules (J)**. There are 1000 joules in 1 kilojoule (kJ). You need between 8000 and 10 000 kJ each day. You need more energy if you are more active and growing quickly.

NUTRITION INFORMATION		
	Per 40 g serving	Per 100 g
Energy	600 kJ (143 kcal)	1500 kJ (358 kcal)
Protein	4.7 g	11.8 g
Carbohydrate	20.9 g	52.3 g
sugars	*10.0 g*	*25.0 g*
starch	*10.9 g*	*27.3 g*
Fat	1.9 g	4.7 g
Calcium	200 mg	500 mg
Iron	27 mg	68 mg
Vitamin C	10 mg	25 mg
Fibre	9.9 g	24.8 g

We also measure energy in calories (cal) or kilocalories (kcal). There are 1000 cal in 1 kcal.

Many food labels show the amounts of different types of carbohydrate.

We measure nutrients and fibre in grams (g) or milligrams (mg). There are 1000 mg in 1 g.

1.11 *Food labels show what is in 100 g of a food and what a normal serving of the food contains. To compare foods you must use the amounts per 100 g.*

Key terms

constipation: when your intestines become blocked.

faeces: solid waste material produced by humans and other animals.

fibre: food substance that cannot be digested but which helps to keep your intestines healthy.

Key term

joule: unit used to measure energy.

4. Look at the food label.
   a) How many grams of protein are in 100 g of the food?
   b) How large is one serving?
   c) How many grams of fat are in one serving?
   d) Give the name of *one* mineral in the food.
   e) How many milligrams of calcium would be in 80 g of this food?
   f) How much energy is in 100 g of the food?
   g) Why does your body need energy?
   h) What process does your body use to release energy from food?

5. Give the name of a nutrient that is a good source of energy.

6. Explain why more active people need to eat more food.

7. **a)** What is 500 mg in grams (g)?

   b) What is 4.7 g in milligrams (mg)?

   c) What is 25 mg in grams?

8. Look at the food label. If you only ate this food, about how much would you need to provide your energy for one day?

Carbohydrates

We need carbohydrates for energy. Respiration releases energy from the carbohydrates in your diet. Starch is the main carbohydrate that we eat. There are many good sources of starch, including rice, pasta, potatoes and bread.

Sugars are carbohydrates found in sweet foods, such as candy and cakes. However, sugary foods may damage teeth.

If your body does not use all the carbohydrates you eat, it turns them into fats and stores them. Overeating carbohydrates makes people get larger. If a person gains a lot of weight, they may become obese.

People with **obesity** are more likely to get **type 2 diabetes**. This disease may damage organs in the body (such as the heart, kidneys and eyes). People with type 2 diabetes should reduce the amounts of carbohydrates and fats they eat and exercise more.

1.12 *These foods are good sources of starch.*

1.13 *You should only have small amounts of sugary foods.*

9. What does your body need carbohydrates for?

10. Name a source of starch in your diet.

11. Give *two* reasons why people should not eat lots of sweet things.

Key terms

obesity: being so overweight that your health is in danger.

type 2 diabetes: disease that may damage organs.

Activity 1.4: Investigating energy in food

How do we compare the energy released by different foods?

Burning releases energy from foods quicker than respiration. Figure 1.14 shows how to heat cold water using burning food. The more energy released, the more the water temperature rises.

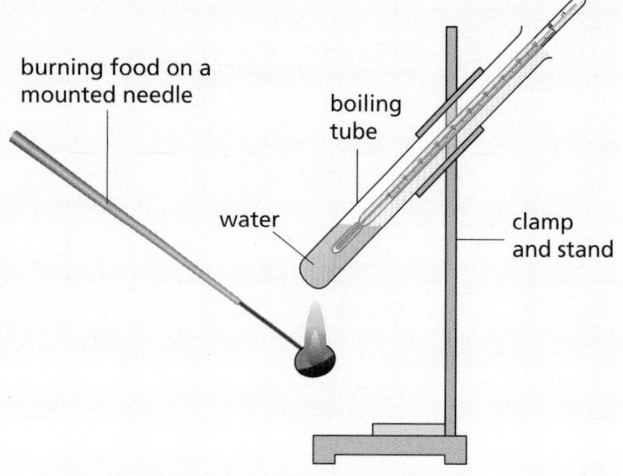

1.14 *Burning food samples to measure temperature rise.*

Here are the instructions for this experiment.

A Add 20 cm^3 of cold water to a boiling tube.

B Measure the temperature of the water.

C Use a balance to find the mass of the food.

D Heat the food until it starts to burn.

E Heat the water using the burning food, until it stops burning.

F Measure the temperature of the water again.

G Calculate the change in the temperature using this equation:

change in temperature = temperature at end – temperature at start

We use this experiment to find the different amounts of energy in foods. The variable we change is the food. The variable we measure is the rise in temperature.

To make a fair comparison between foods, we must keep other variables the same. These are **control variables**, and include the volume of water and the mass of food.

The results of some experiments are in the table.

Food	Temperature of water at start (°C)	Temperature of water at end (°C)	Rise in temperature (°C)
bread	17.2	22.6	
cookie	18.5	28.5	
popcorn	17.9	25.8	

A1 Copy the table and calculate the rise in temperature for each food.

A2 Which food contained the most energy?

A3 What was used to measure the water temperature?

A4 Suggest *one* way to control risks in this experiment.

A5 Apart from water volume and mass of food, suggest *one* more control variable.

12 Look at figure 1.15. What are the temperatures on thermometers A, B and C?

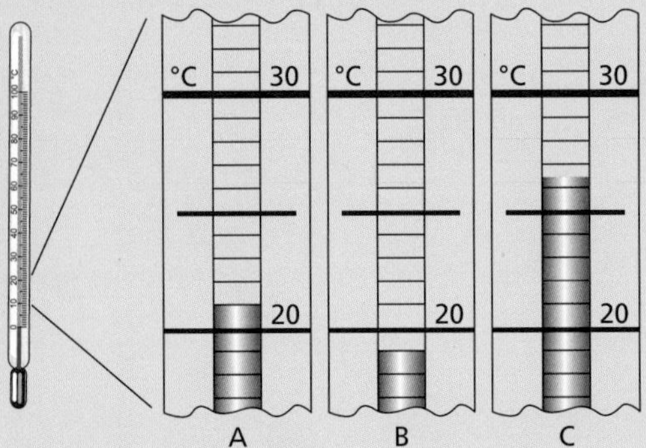

1.15

13 A scientist adds some drops of iodine solution to a slice of cassava. A blue-black colour appears. What does this tell you about cassava?

Key term

control variable: variable that you keep the same during an investigation.

Fats

Your body uses fats (lipids) to store energy. Fat is also stored under your skin to help keep you warm. Good sources of fats include butter, cheese, eggs, oils and milk.

Overeating foods containing fats may make people obese. Fatty foods may also cause problems with a person's circulatory system, and stop blood flowing smoothly.

If you rub a fatty food on some paper, it leaves a greasy mark. This is a test for fats.

1.16 *Some good sources of fats.*

14 Name a source of fats in your diet.

15 A scientist rubs some cheese and some rice on a piece of paper.

a) Predict what the scientist will see.

b) Explain your prediction.

Proteins

Your body uses proteins to make substances to build new cells, repair your body and to let you grow. Muscles are mainly protein and so meats are a good source. Other good sources of protein include beans, cheese, eggs, fish, lentils, milk and tofu.

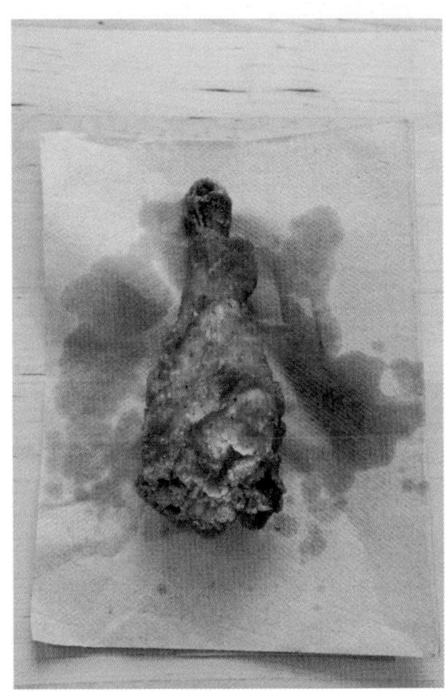

1.17 *A test for fats.*

Your muscles are made of protein. People who need lots of protein in their diets include athletes and body builders, and young people who are growing quickly.

Scientists use Biuret solution to test for proteins. This liquid turns from blue to purple when shaken with food containing proteins.

When scientists test foods for protein using Biuret solution, they see a **trend** (pattern) between the amount of protein and the strength of the purple colour. The more protein, the stronger the purple colour. Scientists usually look for trends in data.

1.18 *Some good sources of proteins.*

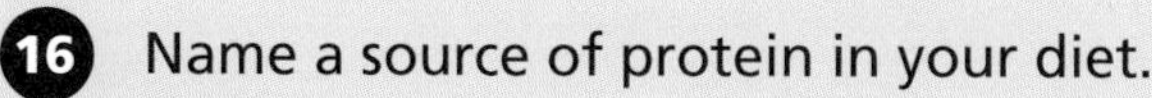

16 Name a source of protein in your diet.

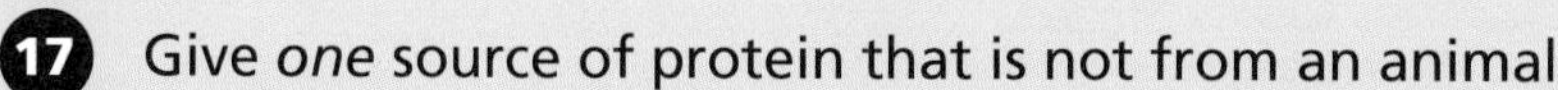

17 Give *one* source of protein that is not from an animal.

18 Explain why an athlete needs to eat lots of protein.

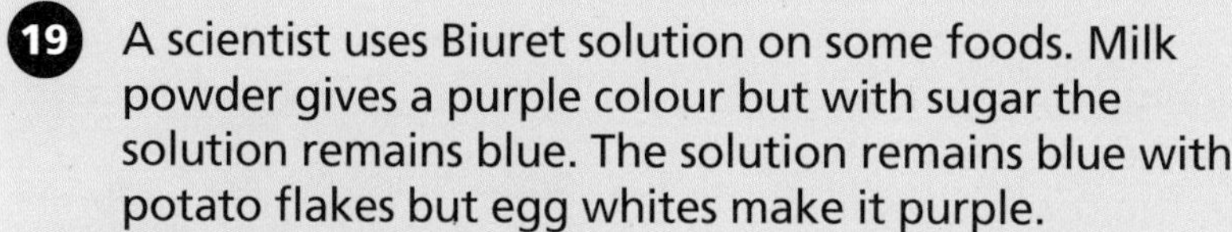

19 A scientist uses Biuret solution on some foods. Milk powder gives a purple colour but with sugar the solution remains blue. The solution remains blue with potato flakes but egg whites make it purple.

a) Present the test results as a table. Include the meanings of the results.

b) Biuret solution will burn your eyes. How would you control the risks when using it?

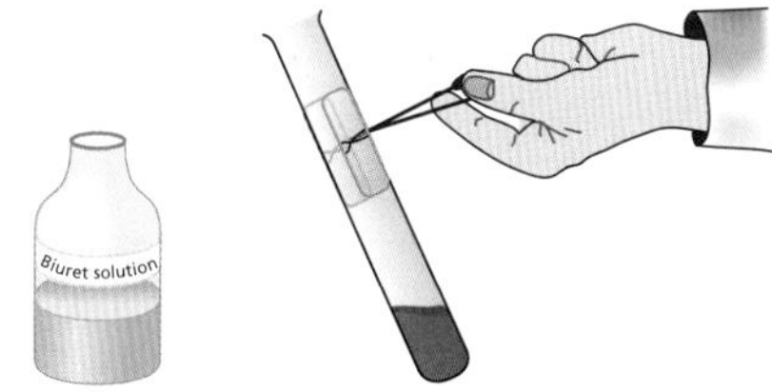

1.19 *The Biuret test for proteins.*

Key term

trend: pattern seen in data.

Vitamins and minerals

You need small quantities of vitamins and minerals in your diet. Each has a function that keeps your body healthy and working properly. Table 1.1 gives some examples.

Vitamin or mineral	Good sources	What it does
vitamin A	carrots, cheese, eggs, oily fish	keeps eyes and skin healthy
vitamin C	lemons, limes, oranges, pineapples	keeps skin and gums healthy
vitamin D	oily fish, red meat, eggs	keeps bones and teeth strong
calcium (a mineral)	bok choy, milk, cheese	used in bones and teeth
iron (a mineral)	beans, meat, dark green vegetables	used in red blood cells

Table 1.1 *Some examples of vitamins and minerals.*

Salt contains the mineral sodium, which nerve cells need to work. However, large amounts of sodium may cause blood to put too much pressure on blood vessels. **High blood pressure** may burst blood vessels and damage organs (such as the heart).

Key term

high blood pressure: when blood puts too much pressure on blood vessels.

20 a) Name a vitamin and a source of that vitamin in your diet.

b) What does the vitamin do?

21 a) Name a mineral and a source of that mineral in your diet.

b) What does the mineral do?

22 Why should you not put too much salt on your food?

Balanced diets

No single food contains all the nutrients you need and so you should eat foods from many sources. If you eat many different foods and get nutrients in the correct amounts, you have a **balanced diet**.

1.20 *For a balanced diet, you should eat lots of fruits and vegetables. You also need foods containing starch. Foods containing lots of fats and sugars should form only a small part of your diet.*

Recommendations for a balanced diet

To help people eat a balanced diet, governments often make recommendations. These recommendations have different names in different countries. Table 1.2 shows some 'Guideline Daily Amounts' for a child in the United Kingdom.

	Amount each day for children aged 5–10
Energy	7500 kJ (1800 kcal)
Protein	24 g
Carbohydrate	220 g
Sugars	85 g
Fat	70 g
Fibre	15 g
Salt	4 g

Table 1.2 *Guideline daily amounts for a child in the UK.*

Key term

balanced diet: eating many different foods to get the correct amounts of nutrients.

23 Table 1.2 recommends a maximum of 85 g of sugars in a day.

a) Name the type of carbohydrate that you should eat more of than sugars.

b) Why should less of the carbohydrate in your diet come from sugars?

24 How does the recommended energy in the table compare to the amount *you* need. Explain your answer.

Activity 1.5: Investigating food recommendations

What are the food recommendations in different countries?

A1 Use different books and/or the internet to discover how much of each nutrient your country recommends. Also discover the values from a country far from yours.

A2 Present your information as a table, to compare the values.

Key facts:

- ✔ Your diet needs to contain nutrients (carbohydrates, fats, proteins, minerals and vitamins) for energy, growth and repair, and health.
- ✔ You need water and fibre in your diet too.
- ✔ Carbohydrates and fats store energy, which is measured in joules (J) or kilojoules (kJ).
- ✔ Eating too much carbohydrate or fat may cause weight gain and make people unhealthy.
- ✔ We use tests to discover which foods contain which nutrients.
- ✔ For a balanced diet you need to eat foods from many sources.

Check your skills progress:

I can identify control variables in investigations.

I can use tables to present and compare information.

I can plan to control risks in investigations.

Malnutrition

Learning outcomes

- To describe what happens if you lack nutrients in your diet
- To explain why a good diet is important for fitness

Starting point

You should know that...	You should be able to...
We need many different nutrients in our diets	Research and present data using secondary sources
Different nutrients have different functions	
Too much of some nutrients may be bad for health	

In many parts of the world, people cannot eat enough of certain nutrients. They may suffer from a **nutritional deficiency** (also called a **deficiency disease**).

A lack of protein causes a nutritional deficiency called **kwashiorkor**. A **symptom** of this disease is that muscles become weak. Fluid collects around the intestines and the muscles are too weak to hold the 'belly' in. People with kwashiorkor often have a 'pot belly'.

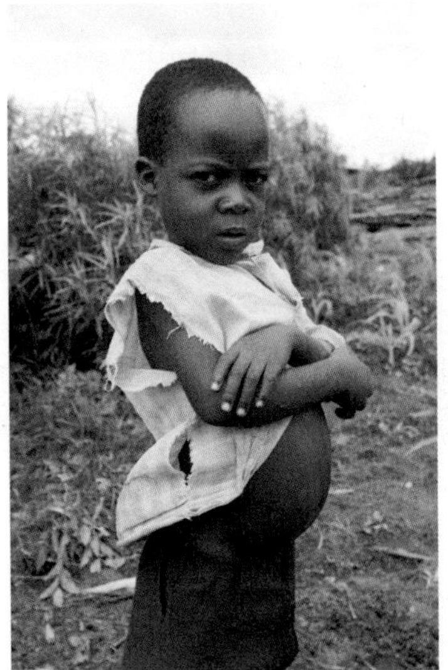

1.21 *A child with kwashiorkor.*

a) What causes kwashiorkor?

b) Explain why muscles become weak in people with kwashiorkor.

A lack of vitamin C causes **scurvy**. A person with scurvy has painful joints and bleeding gums.

Too little vitamin D or calcium causes **rickets**. This causes weak bones, which bend as they grow. Vitamin D is in foods but we also make it in our skins, when exposed to sunlight.

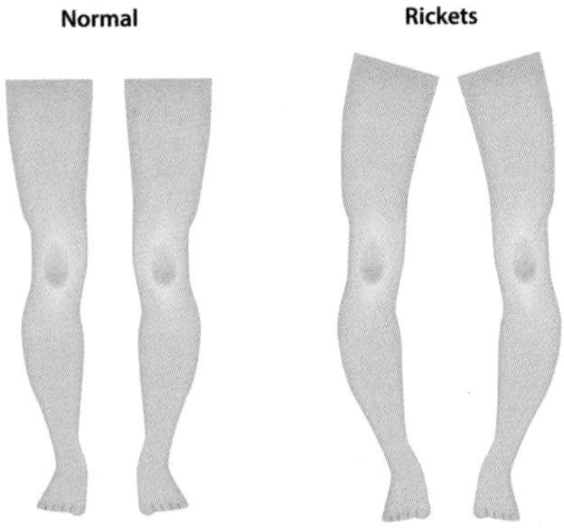

1.22 *Rickets causes curved legs.*

Key terms

deficiency disease: another term for nutritional deficiency.

kwashiorkor: nutritional deficiency caused by a lack of protein.

nutritional deficiency: problem caused by a lack of a nutrient in the diet. Also called a deficiency disease.

symptom: effect of a disease on the body.

2 a) Describe the symptoms of scurvy.

b) Suggest a food that someone with scurvy should eat.

3 a) What is a nutritional deficiency?

b) Name a nutritional deficiency caused by a lack of calcium.

4 Hundreds of children develop rickets in countries in northern Europe each year. Suggest three causes of this.

5 Compare infectious diseases with deficiency diseases. Write a paragraph.

Key terms

rickets: nutritional deficiency caused by a lack of calcium or vitamin D.

scurvy: nutritional deficiency caused by a lack of vitamin C.

Scurvy

Scurvy was a big problem for sailors in the 18th century. A Scottish doctor called James Lind thought that poor diet caused scurvy. He took some sailors and gave them different things to add to their food each day. His results are in table 1.3.

Item added to daily diet	Did the sailor recover from scurvy?
drink of mild acid	No
drink of vinegar	No
drink of salty water	No
eat an orange and a lemon	Yes
drink of grain boiled in water	No

Table 1.3 *Results of Lind's experiment to find a cure for scurvy.*

After Lind's research, scientists began to understand the importance of eating fruits and vegetables. A Hungarian scientist called Albert Szent-Györgyi discovered vitamin C in the 1930s.

6 a) What question did James Lind ask?

b) Make a conclusion from his results.

Activity 1.6: Investigating nutritional deficiencies

What are the causes and symptoms of some nutritional deficiencies?

A1 Use this and other books and/or the internet to learn about problems due to a lack of: vitamin C, vitamin D, iron, calcium and two other vitamins or minerals.

A2 Present your information as a table, to compare the nutritional deficiencies. Include the names of the deficiency diseases, their symptoms and their causes.

Diet and fitness

Fit people easily do everything they need to do each day (such as running upstairs and not being out of breath). Exercise keeps people fit but diet is important too.

People with a problem caused by too much or too little of something in their diets have **malnutrition**. These problems make people less fit and include:

- obesity
- type 2 diabetes
- high blood pressure
- deficiency diseases.

Key term

malnutrition: when a diet contains too much or too little of something, and causes health problems.

7 What sort of malnutrition causes high blood pressure?

8 What is the difference between the causes of high blood pressure and of rickets?

Key facts:

- ✔ Malnutrition is when a diet contains too much or too little of something, which then causes health problems.
- ✔ Malnutrition causes deficiency diseases (such as kwashiorkor and scurvy).
- ✔ Malnutrition also causes obesity, type 2 diabetes and high blood pressure.

Check your skills progress:

I can find and use information from different sources.

I can use tables to present and compare information.

Chapter 1 . Topic 4

Digestion

Learning outcomes

- To identify the organs of the alimentary canal
- To describe the functions of the organs involved in digestion
- To describe what happens in digestion, including the action of enzymes

Starting point

You should know that...	You should be able to...
The food we eat is digested in our bodies by the digestive system	Research and present data using secondary sources
The digestive system contains organs such as the gullet, stomach, small intestine and large intestine	
Food that has been digested goes into the blood	

Food contains many large, complex substances that your body cannot use. Digestion breaks these substances into tiny pieces, which your blood is able to carry.

Digestion occurs in your **alimentary canal** or **gut**. This is made of many different organs that form a tube from your **mouth** to your **anus**.

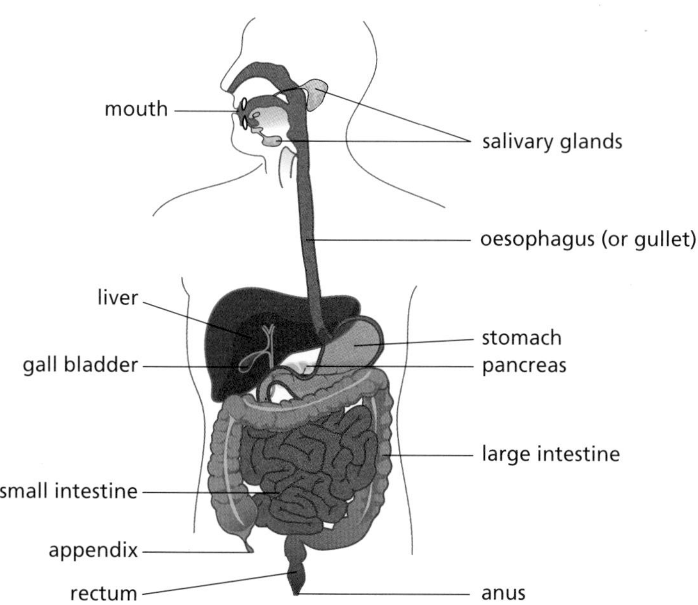

1.23 *The human digestive system contains the organs of the alimentary canal, as well as the salivary glands, liver, pancreas and gall bladder.*

Key terms

alimentary canal: tube that runs from your mouth to your anus.

anus: last organ in the alimentary canal. Faeces leave your body here.

gut: another word for alimentary canal.

mouth: first organ in the alimentary canal.

1. List the organs of the alimentary canal that food goes through.
2. Explain why your body digests food.

In your mouth

When you chew food in your mouth, your teeth grind it into smaller bits. This is **mechanical digestion**. Your **salivary glands** add liquid **saliva**, which makes food easier to swallow.

Saliva is also a **digestive juice** because it breaks apart nutrients using chemical substances, called **enzymes**. This is **chemical digestion**. The enzymes in saliva break down starch and turn it into small sugars.

The smaller pieces of food created by mechanical digestion help chemical digestion to occur. Digestive juices are able to reach more of the food if it is in smaller pieces.

Activity 1.7: Investigating saliva

How does the taste of bread change if you chew it for a long time?

A1 Make a prediction to answer the question above.

A2 Explain your prediction.

3. Describe two functions of saliva.
4. Describe one way in which mechanical digestion is different from chemical digestion.

In your oesophagus

The wall of your **oesophagus (gullet)** contains muscles. These contract and relax to push swallowed food into your **stomach**. This process is **peristalsis**. The organs in the rest of your alimentary canal also use peristalsis to push food along.

Key terms

chemical digestion: digestion that is done by chemical substances.

digestive juice: liquid that contains enzymes to digest food.

enzyme: substance that digests food.

mechanical digestion: digestion that is done by physical actions, such as chewing.

saliva: digestive juice made by salivary glands in your mouth.

salivary gland: organ inside your mouth that makes saliva.

Key terms

gullet: another word for oesophagus.

oesophagus: organ of the alimentary canal. Its muscle walls push food from your mouth into your stomach.

peristalsis: contraction and relaxation of muscles in the alimentary canal that pushes food along.

stomach: organ of the alimentary canal. It makes enzymes to digest proteins and churns food into a smooth soup.

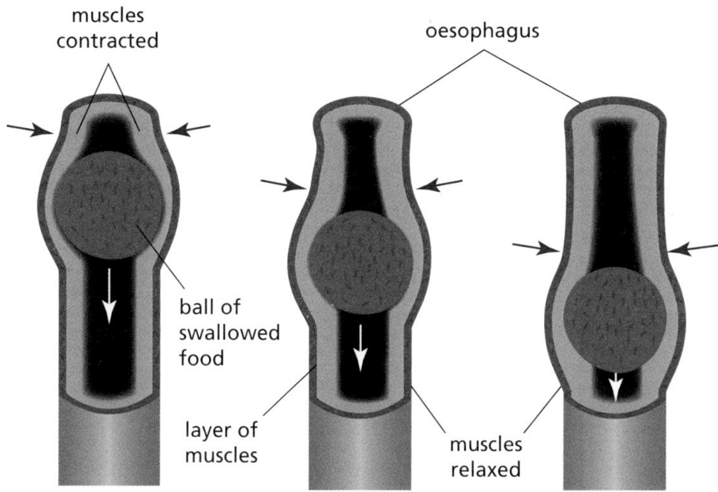

1.24 *Peristalsis moves food through your alimentary canal.*

In your stomach

Your stomach adds digestive juices to attack proteins. Mechanical digestion also occurs, as stomach movements churn and mix the food into a smooth soup.

In your small intestine

Muscles in your stomach push food into your **small intestine**. Your small intestine and **pancreas** add further digestive juices, to digest carbohydrates, proteins and fats.

Your **liver** makes **bile**. This liquid is not a digestive juice but it helps the fat-digesting enzymes to work. Your **gall bladder** stores bile and releases it into your small intestine when needed.

The digested food is now in such tiny pieces that it is able to pass into your blood, through the wall of the small intestine.

5 Which part of the alimentary canal first digests these nutrients?

a) carbohydrates

b) fats

c) proteins

6 **a)** Describe the route bile takes from where it is made into your gut.

b) What is the function of bile?

7 Suggest the name of a substance the small intestine digests starch into.

Key terms

bile: liquid that helps fat-digesting enzymes to work.

gall bladder: organ next to your liver that stores bile.

liver: organ that makes and destroys substances. It makes bile.

pancreas: organ that makes enzymes to digest fats, proteins and carbohydrates.

small intestine: organ of the alimentary canal. It makes enzymes and lets digested food pass into your blood.

In your large intestine

When food reaches your **large intestine**, all that remains are substances that you cannot digest (such as fibre). Your large intestine removes water from the undigested food, to form a solid waste (faeces).

Faeces is stored in your **rectum** and then pushed out through your anus.

Key terms

large intestine: organ of the alimentary canal. It removes water from undigested food to make faeces.

rectum: organ of the alimentary canal. It stores faeces.

Enzyme action

Starch, proteins and fats are made of enormously long chains. During digestion, chemical reactions break these chains into tiny pieces.

A **catalyst** is a substance that speeds up a chemical reaction. Enzymes speed up digestion, and so we call them biological catalysts. We imagine them as tiny scissors, cutting up the chains. When you use a familiar object to think about something complicated, you are using a **model**.

Key terms

catalyst: substance that speeds up a chemical reaction.

model: simple way of showing or explaining a complicated object or idea.

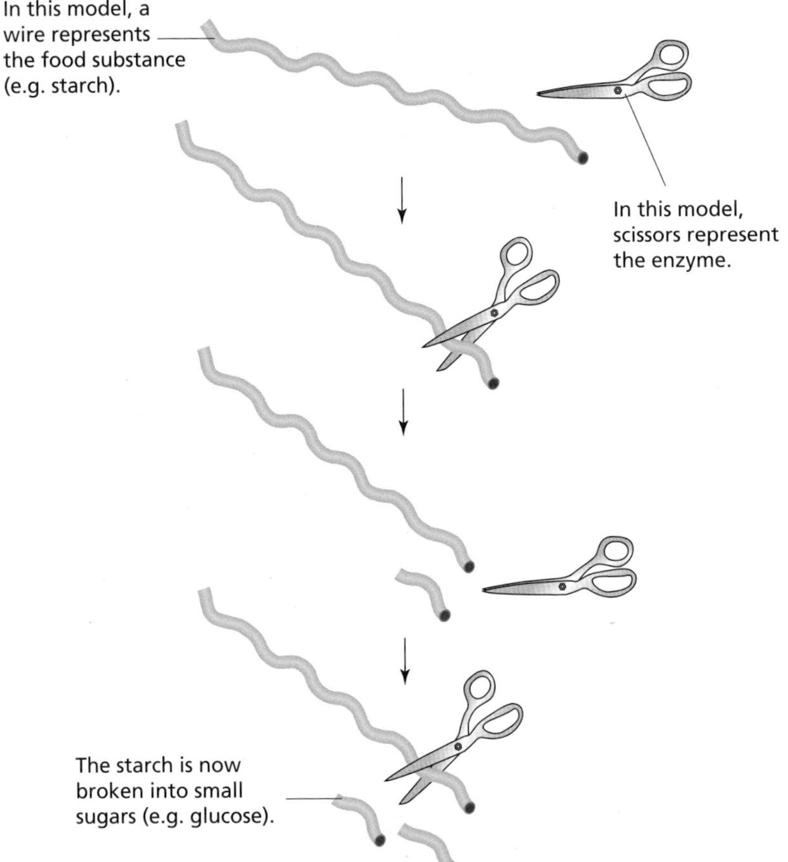
In this model, a wire represents the food substance (e.g. starch).
In this model, scissors represent the enzyme.
The starch is now broken into small sugars (e.g. glucose).

1.25 *We imagine that enzymes are like tiny scissors.*

8. Name the process that moves food through the alimentary canal.
9. Why is there a lot of fibre in the large intestine?
10. Why are enzymes biological catalysts?
11. Use the model of some scissors cutting through string to explain how enzymes work.
12. Draw a flow chart to show what happens in each part of the alimentary canal.

Key facts:

- ✔ You need to digest food for it to be useful inside your body.
- ✔ The digestive system contains the alimentary canal, and some other organs (such as the liver, pancreas and salivary glands).
- ✔ The alimentary canal is a series of organs that form a tube from your mouth to your anus.
- ✔ Digestive juices contain enzymes, which act as biological catalysts to digest food.

Check your skills progress:

I can make predictions based on scientific knowledge.

End of chapter review

Quick questions

1. A lack of fibre in your diet can cause:

   a anaemia   b rickets   c scurvy   d constipation [1]

2. An example of a biological catalyst is:

   a an enzyme   b a cell   c light   d water [1]

3. Digestion starts in your:

   a liver   b small intestine   c mouth   d stomach [1]

4. Food is pushed through your gut by:

   a diffusion   b peristalsis   c peritonitis   d perfusion [1]

5. Bile is stored in your:

   a liver   b gall bladder   c stomach   d pancreas [1]

6. Obesity is caused by a type of malnutrition.

   (a) What is obesity? [1]

   (b) Describe the type of malnutrition that causes it. [1]

7. **(a)** Give *one* reason why we need to eat each of these nutrients:

   - carbohydrates
   - fats
   - proteins. [3]

   (b) Name *one* other nutrient that we need. [1]

   (c) Describe what may happen if we do not eat enough of that nutrient. [1]

8. A student is looking at some pondweed, in a large beaker of water. A lamp is shining on the plant and bubbles are rising from its leaves.

   (a) What gas do the bubbles contain? [1]

   (b) Through which part of the leaf does the gas leave the plant? [1]

   (c) Why is the plant making these bubbles? [1]

   (d) What raw materials does the plant need for this process? [1]

9. Palisade cells in a leaf are adapted for photosynthesis.

   (a) Palisade cells contain a lot of a green substance. Name this substance. **[1]**

   (b) In which parts of the cell is the green substance found? **[1]**

   (c) Describe the function of the green substance. **[1]**

10. Food moves from your oesophagus to your large intestine.

   (a) Which *two* organs does food pass through on this journey? **[2]**

   (b) Explain why the food in your large intestine is mainly fibre. **[1]**

   (c) What is the function of your large intestine? **[1]**

   (d) Where in your alimentary canal is digested food absorbed into your blood? **[1]**

Connect your understanding

11. Scientists use chemical tests to find out about the substances in things. Figure 1.26 shows the test for a nutrient on an unripe banana.

1.26 *Testing a banana.*

   (a) Name the liquid that is being added to the banana. **[1]**

   (b) Describe the colour change as the liquid touches the banana. **[1]**

   (c) What does this colour change tell you about the banana? **[1]**

   (d) Some other tests are shown in the table. Copy and complete the table. **[7]**

Test	What substance is being tested for?	Positive result	Negative result
Add Biuret solution to food		solution turns purple	solution stays blue
Rub food on a piece of paper			no greasy mark on paper
Place a glowing splint in a gas		splint relights	
Add limewater to a gas			limewater stays clear

12. Figure 1.27 shows a plant leaf with some foil covering part of it. The plant grows under a bright light. A student tests the leaf to see if it contains glucose.

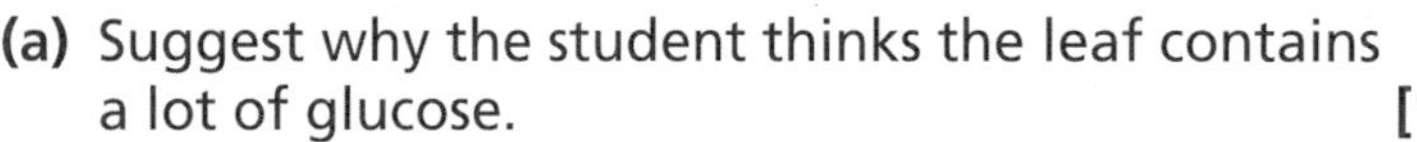

(a) Suggest why the student thinks the leaf contains a lot of glucose. **[1]**

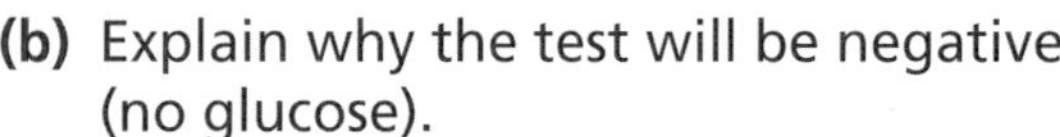

(b) Explain why the test will be negative (no glucose). **[1]**

1.27

(c) What substance should you test for? **[1]**

(d) In which areas of the leaf will you find this substance? Choose *one* or more of A, B and C. **[1]**

(e) Explain your answer to part (d). **[1]**

13. A scientist uses some burning foods to heat up some water. She wants to know which food contains the most energy.

Food	**Temperature of water at start (°C)**	**Temperature of water at end (°C)**
Dried potato	20.1	23.0
Cookie	20.1	24.8
Popcorn	20.1	24.2

(a) Which variable (shown in the table) did she control? **[1]**

(b) Suggest one other control variable. **[1]**

(c) Which food contained the most energy? **[1]**

(d) Explain your answer to part **c)**. **[1]**

(e) What units is energy measured in? **[1]**

14. Unripe apples may taste a bit like potatoes because they contain starch. As they ripen, enzymes break down the starch and the apples become sweeter.

(a) What does the enzyme break the starch into? **[1]**

(b) Explain why enzymes are catalysts. **[1]**

Challenge questions

15. The table shows some information about different foods in a meal.

Values per serving	**Beans**	**Avocado**	**Rice**
Energy	500 kJ	700 kJ	850 kJ
Protein	16 g	2g	4.6 g
Carbohydrate (including sugars)	4 g (4 g)	0.8 g (0.7 g)	48 g (0.3 g)
Fat	2 g	29 g	3.6 g
Calcium	20 mg	0.024 g	2.5 mg
Iron	–	1 mg	4 mg
Fibre	–	7 g	0.6 g

(a) A scientist says "This meal is balanced." Suggest what she means. **[1]**

(b) A student says "This shows that rice contains more energy than beans or avocados." Explain why the student cannot say this. **[1]**

(c) Calculate the number of milligrams of calcium in this serving of avocado. **[1]**

2

Chapter 2

Respiration and circulation

What's it all about?

During an eye test, an eye doctor looks at tiny blood vessels inside your eye. The optician may take a photo, like this one. All cells, including cells in your eyes, need a constant supply of oxygen and food from blood. Your heart pumps blood around your body by putting pressure on it. However, too much pressure can damage the delicate blood vessels in the eye. This can cause eye problems.

You will learn about:

- The parts and function of the human circulatory system
- Aerobic respiration
- The parts and function of the human respiratory system
- How gases are exchanged between your blood and the air
- The effects of smoking
- How water and mineral salts are absorbed and transported in flowering plants

You will build your skills in:

- Developing questions that can be investigated
- Collecting evidence
- Using creative thinking to develop explanations
- Planning investigations to test ideas

Human circulatory system

Learning outcomes

- To describe the parts and functions of the circulatory system
- To explain how oxygen and food substances are delivered to all cells
- To describe some circulatory system disorders

Starting point

You should know that...	You should be able to...
The circulatory system is a group of organs that gets blood around your body	Explain how scientists think up ideas, make predictions, and collect evidence to test their ideas
Blood is carried around your body in blood vessels	Use tables, bar charts and line graphs
Blood contains specialised cells	
Blood transports substances, including oxygen	
Poor diet can harm the circulatory system	

Circulation

Your **heart** is a pump that pushes **blood** around your body through tubes called **blood vessels**. This movement of blood is your **circulation**. Your heart, blood and blood vessels form your **circulatory system**. It supplies all your cells with the substances they need.

All cells need oxygen, which enters blood in your lungs. When we model the circulatory system using diagrams, blood that contains lots of oxygen is coloured red. Blood with only a little oxygen is coloured blue. In real life, it is a dark brown-red colour. Using blue in diagrams makes the difference more obvious.

1. Which life process do your cells need oxygen for?
2. List *two* organs in your circulatory system.
3. Evaluate the advantages and disadvantages of using blue in diagrams to show blood with only a little oxygen.

The heart

There are four **chambers** inside your heart. You have two **atria** at the top of your heart – your left atrium and your right atrium. Beneath these are your **ventricles**.

Key terms

atria: chambers at the top of your heart. You have a left atrium and a right atrium.

blood: liquid organ that carries substances around the body.

blood vessels: tube-shaped organs that carry blood around the body.

chamber: space inside the heart that fills with blood and empties again.

circulation: movement of blood around the body.

circulatory system: group of organs that gets blood around the body.

heart: organ that pumps blood through blood vessels.

ventricle: chamber at the bottom of your heart. You have a left ventricle and a right ventricle.

Your heart is made of different tissues, including muscle **tissue**, nerve tissue and some fat tissue. The muscle tissue **contracts** and **relaxes** to make the chambers smaller and larger. This pumps blood through your heart.

Each contraction of your heart is a **heartbeat**. The number of times this happens in one minute is your **heart rate**. We measure heart rate in 'beats per minute' (bpm).

Key terms

tissue: group of cells of the same type.

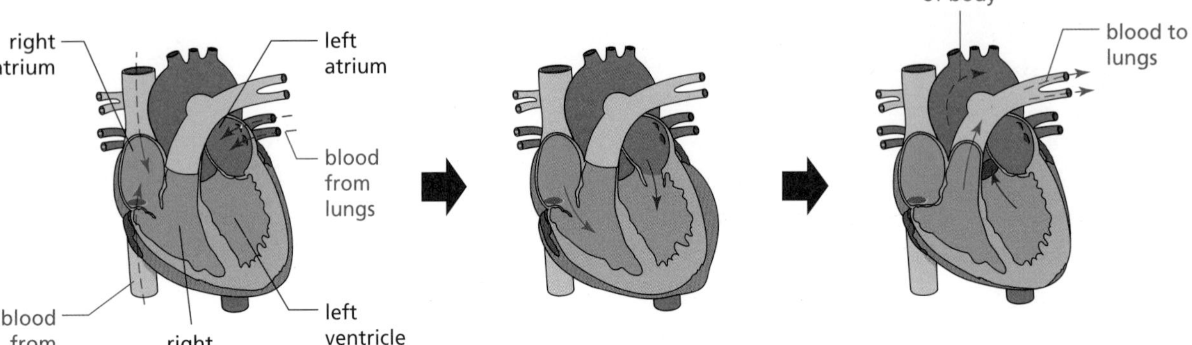

right atrium
left atrium
blood from lungs
blood from body
right ventricle
left ventricle
blood to rest of body
blood to lungs
1 The muscles in the walls of the atria are relaxed. Blood flows into the atria, from blood vessels.
2 The atria muscles contract. This pushes blood into the ventricles.
3 The ventricle muscles contract. This pushes blood into blood vessels leading out of the heart. The atria fill up and the process starts again.

2.1 *Muscle tissue contracts and relaxes to push blood through your heart. We draw a heart as though it belongs to someone facing us. The left ventricle and left atrium appear on the right of the drawing.*

4 Look at figure 2.2.

a) Name the chamber at the bottom of the photo.

b) Name the chamber at the top of the photo.

c) Most of the tissue is a red colour. Give the name of this tissue.

d) Explain how this tissue pushes blood out of a heart chamber.

5 Which side of the heart contains blood with less oxygen in it? Explain your reasoning.

6 Jayden's heart beats 40 times in 30 seconds. Calculate his heart rate.

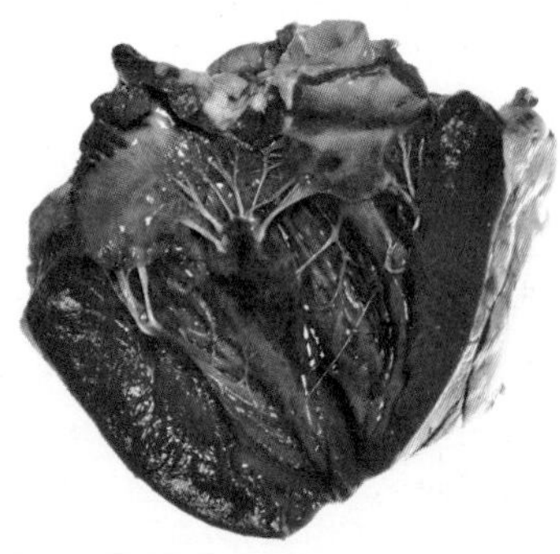

2.2 *Inside a mammal's heart.*

Stethoscopes

Doctors listen to a patient's heart using a stethoscope. A French doctor, René Laënnec, invented the first stethoscope in 1816. It was made of 24 pieces of paper rolled into a tube but later ones were wooden tubes. In 1851, an Irish doctor, Arthur Leared, invented a flexible stethoscope with two earpieces. This is the type that doctors use today.

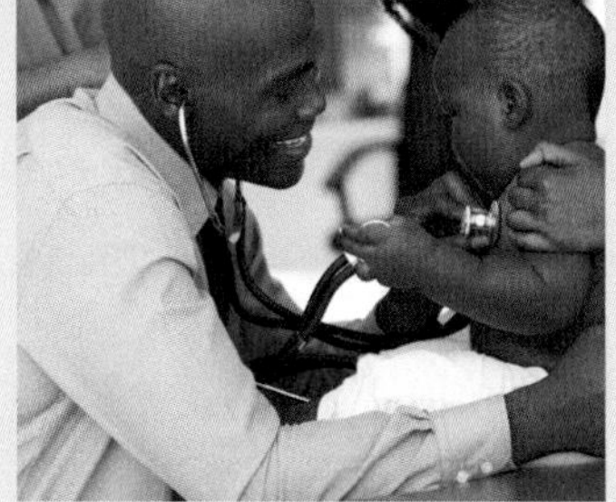

2.3 *A modern stethoscope.*

Another type of stethoscope is the Pinard horn. Doctors and midwives use them to listen to the heartbeats of babies still inside their mothers. A French doctor, Adolphe Pinard, invented it in 1895.

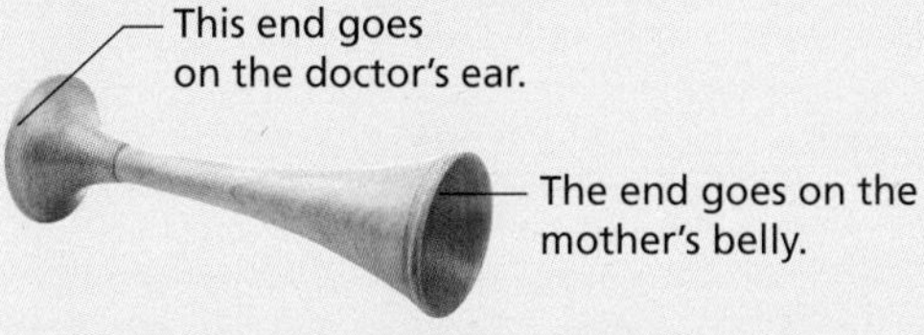
This end goes on the doctor's ear.
The end goes on the mother's belly.

2.4 *A Pinard horn is hollow and made of two cones.*

Key terms

contract (muscle): when muscle tissue gets shorter and fatter, it contracts.

heartbeat: squeezing of the muscles in the heart wall to push blood into blood vessels.

heart rate: the number of heartbeats in one minute.

relax (muscle): when muscle tissue stops contracting, it relaxes.

Activity 2.1: Investigating stethoscopes

How do we hear a heartbeat?

Design and make a simple stethoscope so that you can listen to a friend's heartbeat.

A1 Describe what the heartbeat sounds like.

A2 Using a watch and your stethoscope, discover your friend's heart rate.

Blood vessels

Your circulatory system has two circuits. One takes blood from the heart to the lungs and back to the heart. The other takes blood from the heart to the rest of the body and back again. A circulatory system with two circuits is a **double circulatory system**.

Altogether, your heart pumps about 5 litres of blood every minute, through about 100 000 km of blood vessels!

Key term

double circulatory system: circulatory system in which the heart pumps blood around two circuits. In humans, one circuit supplies the lungs, and the other supplies the rest of the body.

The blood vessels that leave your heart are arteries. An **artery** from the right ventricle takes blood to the lungs. The **aorta** is a large artery that carries blood from the left ventricle towards the rest of the body.

Blood vessels that carry blood back to your heart are **veins**.

Thousands of **capillaries** form fine networks, which connect arteries to veins. These networks are in every tissue, and ensure that all your cells have a supply of blood.

Key terms

aorta: large artery that leaves the left ventricle of your heart.

artery: thick-walled blood vessel that carries blood away from the heart.

capillary: tiny blood vessel that carries blood from arteries to veins.

vein: thin-walled blood vessel that carries blood towards the heart.

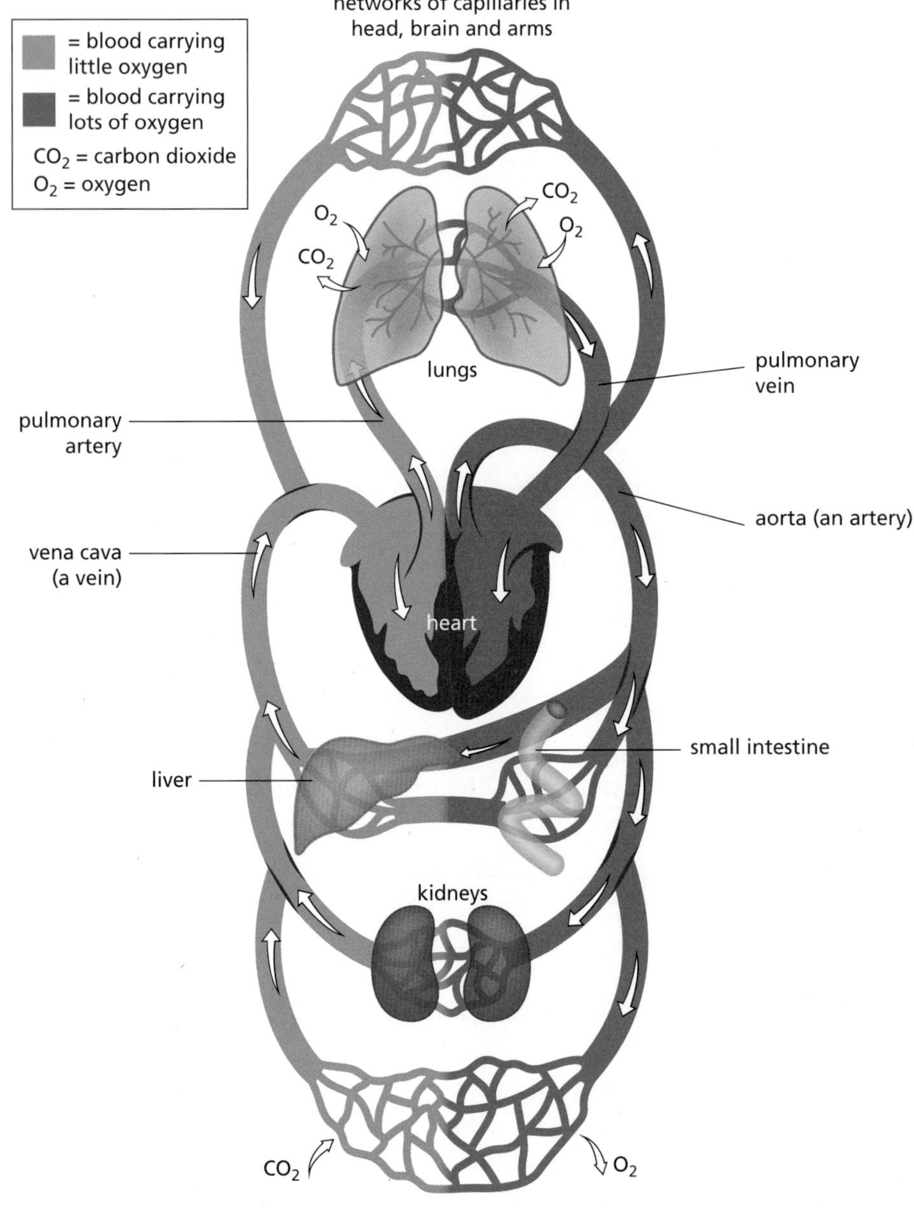

2.5 *Blood vessels have many branches and form many networks of capillaries. This diagram shows only a few of them.*

7 **a)** Name the *three* different types of blood vessel.

b) In which order does blood flow through these, after it leaves the heart?

8 Look at figure 2.5.

a) Give the name of the blood vessel that takes blood to the lungs.

b) Give the name of a vein that returns blood into the heart.

c) Why are capillaries arranged into fine networks?

9 Humans have a double circulatory system. What does this mean?

10 Look at figure 2.5.

a) What is different about the arteries carrying blood to the lungs, compared with those carrying blood to the rest of the body?

b) The muscle wall of the left ventricle is much thicker than the right ventricle. Suggest an explanation for this.

Capillaries

Capillaries are adapted for their function. They have walls that are only one cell thick, and there are gaps between the cells. This allows substances to enter and leave capillaries easily.

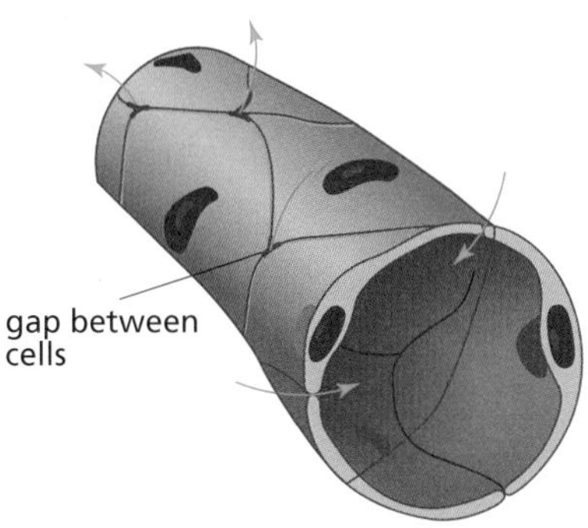

2.6 *Capillary walls allow substances to enter and leave easily.*

Arteries

The heart pushes blood out under high pressure. Artery walls are thick and strong, so that this pressure does not break them. They also have a layer of elastic fibres, which lets them stretch when blood enters. A wave of stretching then passes along an artery, which you feel as a **pulse**. Your pulse is *not* your blood moving.

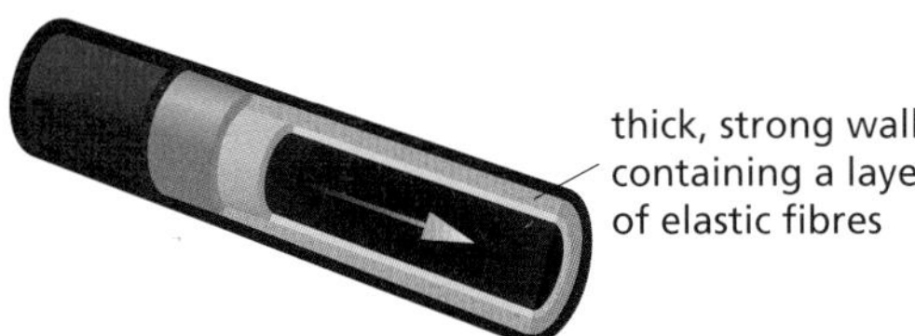

2.7 *Arteries are adapted for their function.*

Key term

pulse: wave of stretching along the wall of an artery each time the heart beats.

Activity 2.2: Measuring your pulse rate

How can you measure your pulse rate?

Your pulse rate is the number of times you feel your pulse in a minute, in 'beats per minute' (bpm). To feel your pulse, press two fingers on your wrist as shown.

A1 Work out your pulse rate. You may find it easier to count the pulses you feel in 15 or 30 seconds and then do a simple calculation.

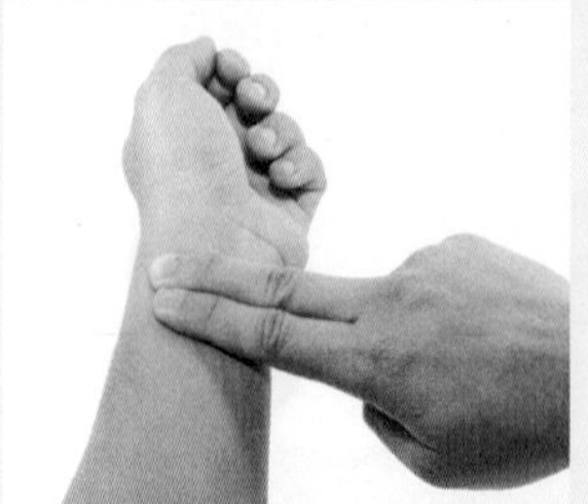

2.8 *Finding your pulse.*

11 Why is it useful for substances to be able to leave capillaries easily?

12 A student feels 15 pulse beats in 10 seconds. Calculate the pulse rate.

13 **a)** State *one* similarity between a heartbeat rate and a pulse rate.

b) State *one* difference between the two.

Veins

Veins do not have thick walls because the blood in them is only under low pressure. However, low pressure means that blood may only flow slowly. To help push the blood along, when muscles move your arms and legs they also squeeze the veins. The veins contain **valves**, which shut if blood starts flowing in the wrong direction.

Key term

valve: flaps of tissue that only allow blood to flow in one direction.

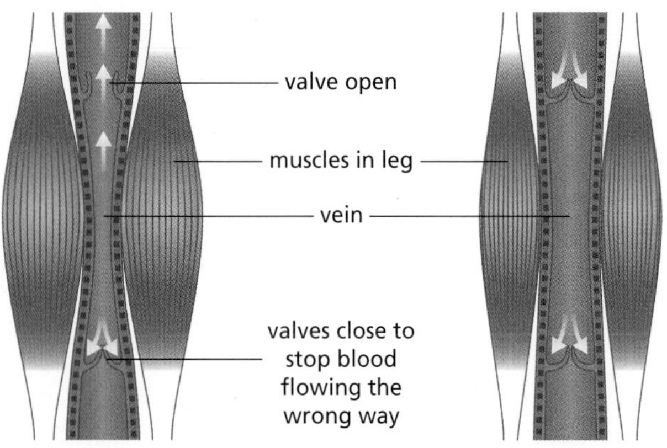

2.9 *Muscles attached to your skeleton help to push blood along veins. Veins contain valves to stop blood flowing in the wrong direction.*

14 Why do veins *not* need thick walls?

15 Draw a table to compare the structures and functions of the *three* types of blood vessel.

16 Look carefully at figure 2.1. Explain the reason for the flaps of tissue between the atria and the ventricles.

Activity 2.3: Investigating the discovery of the circulatory system

The following statements show the discoveries made by different scientists around the world, as people tried to understand how the circulatory system works.

A1 Write the statements in the order in which they happened.

A2 Do some research to check your answers.

A3 Find out about one other scientist who contributed to our understanding of the human circulatory system. Write a paragraph about what he or she did.

- William Harvey: An English doctor. He showed how valves in veins stopped blood flowing the wrong way. He predicted the existence of tiny blood vessels, linking arteries to veins.
- Galen: A Roman doctor. He showed that arteries contained blood. He thought that blood flowed through tiny holes in the heart from the right ventricle to the left ventricle.
- Erasistratus: A Greek doctor. When he dissected bodies, he found air in arteries and concluded that arteries carried air around the body.
- Ibn-al-Nafis: An Islamic doctor, born in modern-day Syria. He showed there were no holes in the middle of the heart, and thought that blood flowed from the right side of the heart to the lungs and then back to the left side. From here it was pumped around the body.
- Marcello Malpighi: An Italian doctor. He discovered capillaries, using a microscope.

Blood

About half your blood is a pale-yellow liquid called **plasma**. It contains many dissolved substances, including:

- digested food substances (absorbed from the small intestine)
- **urea** (a waste product made in the liver, which your kidneys **excrete**)
- carbon dioxide (a waste product of **respiration**, which your lungs excrete).

Key terms

excrete: getting rid of wastes made inside an organism.

plasma: liquid part of the blood.

respiration: chemical process that happens in all parts of an organism to release energy.

urea: waste product made by the liver and excreted by the kidneys.

17 Suggest the name of *one* digested food substance in plasma.

18 From where to where does plasma carry the following substances?

a) urea

b) carbon dioxide

c) digested food substances

19 a) Sketch a diagram to model the circulatory system. Use boxes to represent the organs and capillary networks.

b) Add labels to explain what your model shows.

c) Add labelled arrows to show *one* route taken by urea, carbon dioxide and digested food substances. Use a different colour for each substance.

20 What liquid do mammals produce, which contains a lot of urea?

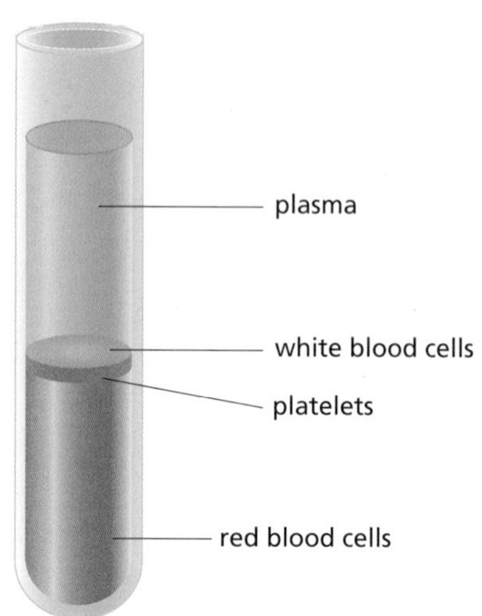

2.10 *Blood contains many different parts.*

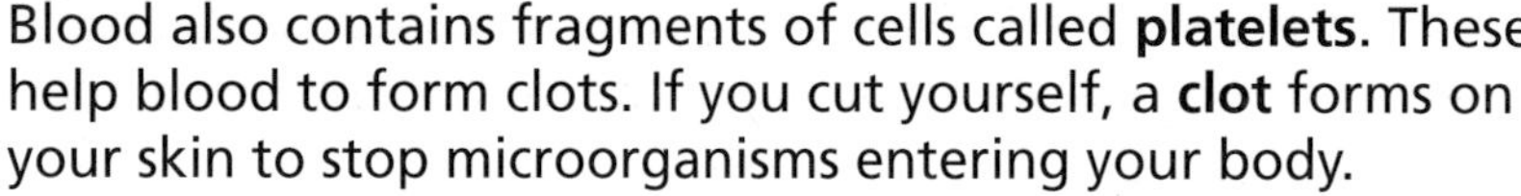

Blood also contains fragments of cells called **platelets**. These help blood to form clots. If you cut yourself, a **clot** forms on your skin to stop microorganisms entering your body.

Red blood cells contain **haemoglobin**. This is a protein, which collects oxygen in your lungs. The haemoglobin releases the oxygen to the cells in your tissues.

Red blood cells do not contain nuclei, and so there is more space for haemoglobin. They have a dimpled shape, which gives them a large **surface area** for their size. The larger the surface area, the faster substances can enter and leave a cell.

White blood cells protect your body from infections and are able to change shape easily. This lets them squeeze through the gaps in capillary walls and get into tissues. Some white blood cells wrap themselves around microorganisms and destroy them.

21 Which part gives blood its colour?

22 Draw a table to show what the different parts of the blood do.

23 Describe, fully, how you could use a micropscope to observe some blood cells.

Key terms

clot: thick mass of blood cells, stuck together.

haemoglobin: substance that collects oxygen.

platelet: cell fragment that helps your blood to clot.

red blood cell: cell that contains haemoglobin so it can carry oxygen.

surface area: the area of a surface, measured in squared units such as square centimetres (cm^2).

white blood cell: cell that helps destroy microorganisms.

 a) How are red blood cells adapted to contain as much haemoglobin as possible?

b) Why is this useful for red blood cells?

 Compared to the number of red blood cells, how many white blood cells are in blood? Explain how you reached your answer.

Activity 2.4: Investigating changes in pulse rate

How does your pulse rate change during exercise?

When you exercise, your muscle cells use more substances from the blood (for respiration). In this activity, you will investigate whether exercise affects pulse rate.

A1 Write down a scientific question that you want to answer.

A2 Plan an investigation method to answer your question. Make sure you include the variable you will change, the variable you will measure and any variables you will keep the same (control variables).

A3 State at least *one* way to control risks.

A4 Predict what will happen and explain your prediction.

A5 Present your results neatly.

A6 Compare your prediction with your results, and make a conclusion.

Cell: red blood cell
Function: carries oxygen
Adaptations:
- no nucleus so that there is more room for haemoglobin
- its indented shape increases its surface area, so that it can absorb oxygen quickly

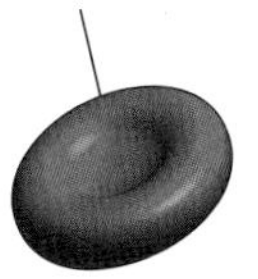

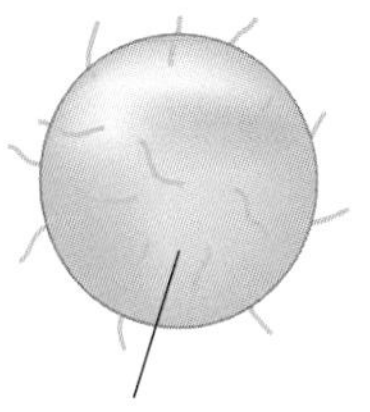

Cell: white blood cell
Function: to destroy microorganisms
Adaptation: flexible shape allows it to squeeze into all different parts of the body.

2.11 *Blood cells are adapted for their functions.*

Circulatory system disorders

Fatty substances may form plaques inside arteries. These make arteries narrower. This happens more often in people who eat too much fat, take no exercise or smoke. Very narrow arteries stop tissues getting enough food and oxygen from blood.

A heart must pump harder to push blood through narrower arteries. This increases the pressure of the blood. High blood pressure may damage the heart, eyes and kidneys.

Sometimes a blood clot forms on a **plaque**. This can stop blood flowing. The clot may also break off the plaque and stick in another artery, where it stops blood flow.

Heart muscle starts to die if it does not get enough blood. This is a **heart attack**. Nerve tissue in the brain will also die if it does not get enough blood. This is a **stroke**. Heart attacks and strokes may both kill people.

Key terms

heart attack: when heart muscle cells start to die and the heart does not pump properly.

plaque: lump of fatty material that builds up inside an artery.

stroke: when brain cells die due to a lack of blood (which is caused by a blocked blood vessel in the brain).

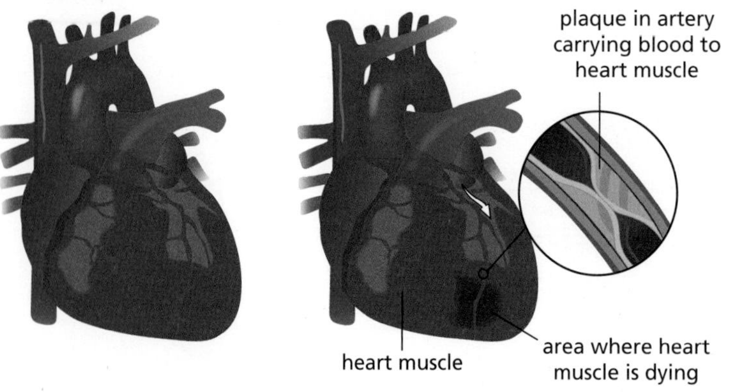

2.12 *A comparison of the appearance of a healthy and unhealthy heart.*

26 Explain why heart muscle tissue needs a supply of blood.

27 Explain what happens if nerve tissue in the brain does not get enough blood.

Activity 2.5: Investigating circulatory system disorders (problems)

How does someone's lifestyle affect their circulatory system?

You are going to prepare some advice for people in your area about how to prevent circulatory system problems. Use this and other books, the internet or information from a local health centre.

A1 Choose *one* circulatory system disorder (such as high blood pressure, coronary heart disease, stroke).

A2 Prepare a way of telling others about this disorder. This could be a poster, a leaflet or a presentation.

- Describe the disorder and its symptoms.
- Explain what causes the disorder.
- Include some ways of preventing this disorder.

Key facts:

- ✔ The human circulatory system contains the heart, blood and blood vessels (arteries, veins, capillaries).
- ✔ The circulatory system ensures that all cells have enough oxygen and food, and removes waste products.
- ✔ Blood plasma carries dissolved food substances, and waste products (such as urea and carbon dioxide).
- ✔ Red blood cells carry oxygen.
- ✔ White blood cells destroy microorganisms.
- ✔ Platelets help the blood to form clots.
- ✔ Disorders of the circulatory system occur when fatty plaques make arteries so narrow that tissues do not get enough blood.

Check your skills progress:

I can plan investigations to test ideas.

I can identify variables in investigations.

I can make and explain predictions.

I can present results using tables.

Human respiratory system

Learning outcomes

- To model aerobic respiration using a word equation
- To explain how oxygen enters the blood, and how carbon dioxide is removed
- To describe the parts and functions of the respiratory system

Starting point

You should know that...	You should be able to...
The respiratory system is a group of organs that get oxygen into the blood and remove carbon dioxide	Explain how scientists think up ideas, make predictions, and collect evidence to test their ideas
All cells respire	Use tables, bar charts and line graphs
Respiration requires oxygen and produces carbon dioxide	
Oxygen is carried by red blood cells	

Aerobic respiration

All cells release energy from glucose (a sugar we get by digesting carbohydrates). This process also needs oxygen from the air and so we call it **aerobic respiration**.

Aerobic respiration is a chemical reaction. We model it using a **word equation**.

oxygen + glucose → carbon dioxide + water

The reactants are on the left of the arrow, and the products are on the right.

Your **respiratory system** (or 'breathing system'), your circulatory system and your digestive system all work together, to ensure that all the cells in your body get the substances they need for aerobic respiration.

a) What are the reactants in aerobic respiration?

b) Explain how these reactants get from outside the body to the cells.

Gaseous exchange

Oxygen enters your blood in your lungs. At the same time, carbon dioxide leaves your blood. This exchange of gases is **gaseous exchange**.

Key terms

aerobic respiration: respiration that requires oxygen to release energy from glucose.

gaseous exchange: when two or more gases move from place to place in opposite directions.

respiratory system: group of organs that get oxygen into the blood and remove carbon dioxide.

word equation: model showing what happens in a chemical reaction, with reactants on the left of an arrow and products on the right.

Oxygen exists as tiny **molecules**, which are constantly moving. If there are many oxygen molecules in one place and fewer in another, there is an overall movement of molecules towards the place with fewer molecules. We call this **diffusion**.

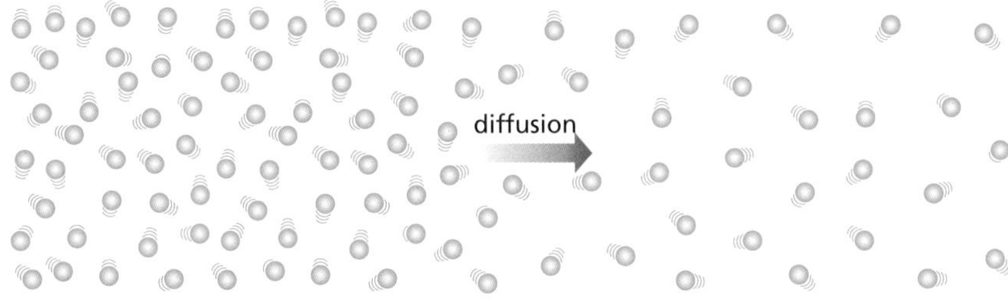
diffusion

2.13 *Diffusion of molecules.*

The lungs contain millions of pockets, called alveoli. Capillaries cover each **alveolus**. There are more oxygen molecules inside an alveolus than in the blood in a capillary. This means that oxygen diffuses from the alveolus into the blood.

There are more carbon dioxide molecules in the blood than in an alveolus. So, carbon dioxide diffuses from the blood into the alveolus.

Key terms

alveolus: tiny, pocket-shaped structure in lungs where gaseous exchange happens. The plural is alveoli.

diffusion: the spreading out of particles from where there are many (high concentration) to where there are fewer (lower concentration).

molecule: group of two or more atoms joined together. Oxygen, carbon dioxide and water all exist as molecules.

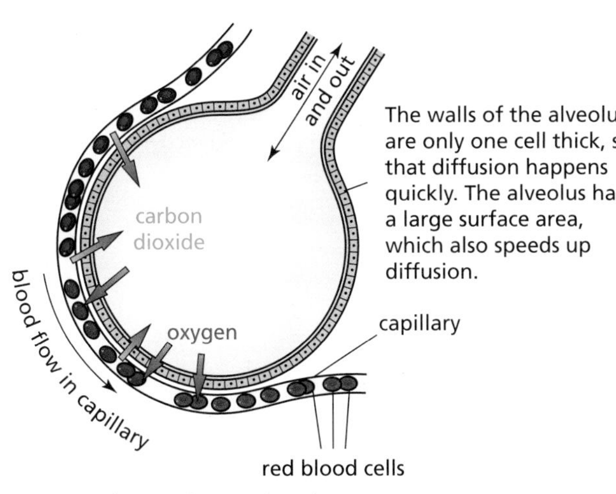
air in and out
The walls of the alveolus are only one cell thick, so that diffusion happens quickly. The alveolus has a large surface area, which also speeds up diffusion.
carbon dioxide
oxygen
capillary
blood flow in capillary
red blood cells

2.14 *Gas exchange in an alveolus.*

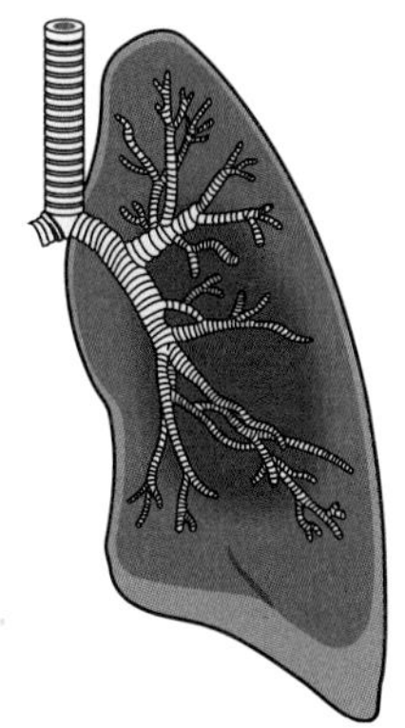

2.15 *The inside of a lung is full of alveoli.*

2. Describe gaseous exchange in the lungs.
3. Explain why carbon dioxide molecules move from the blood into an alveolus.
4. Look at figure 2.15. Explain why the lung looks like a sponge inside.
5. Why does an alveolus have a thin wall?

6 a) How do alveoli change the surface area of the inside of a lung?

b) Why is this adaptation important?

7 Explain how capillaries are adapted for gaseous exchange.

Parts of the respiratory system

Your respiratory system moves air in and out of your **lungs**, through a series of tubes. The largest tube is the **trachea**, which is divided into two bronchi. Each **bronchus** goes into a lung, and leads to many smaller tubes called **bronchioles**.

Other parts of the respiratory system include the **ribs** and **diaphragm**.

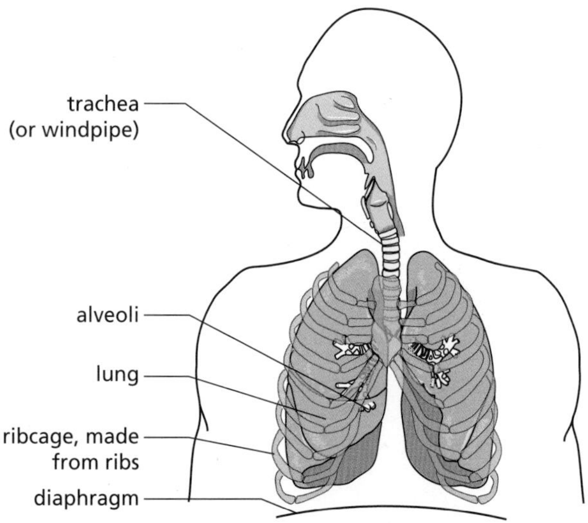

2.16 *The parts of the respiratory system. The area inside your* ***ribcage*** *is your* ***chest****. Your lungs do not join to any part of your ribcage or your chest.*

8 Give another name for the windpipe.

9 The trachea divides into many smaller tubes. What structures are at the ends of all these tubes?

Key terms

bronchioles: small tubes leading from the bronchus in a lung.

bronchus: large tube leading from the trachea into a lung. Plural is bronchi.

chest: area inside the body between the ribcage, neck, backbone and diaphragm.

diaphragm: organ that helps breathing.

lungs: organs that get oxygen into the blood and remove carbon dioxide.

rib: bone that helps to protect your heart and lungs.

ribcage: all your ribs.

trachea: tube-shaped organ that allows air to flow in and out of your lungs.

When you **inhale** (breathe in), muscles between your ribs contract and move your ribcage upwards and outwards. Muscles in your diaphragm contract and flatten it. This increases the volume of your chest. Air flows through your nose into your lungs.

When you **exhale** (breathe out), the muscles relax. Your ribcage falls, and your diaphragm rises. Your chest volume gets smaller and air flows out of your lungs.

The number of times you inhale and exhale in one minute is your **breathing rate** (in breaths per minute).

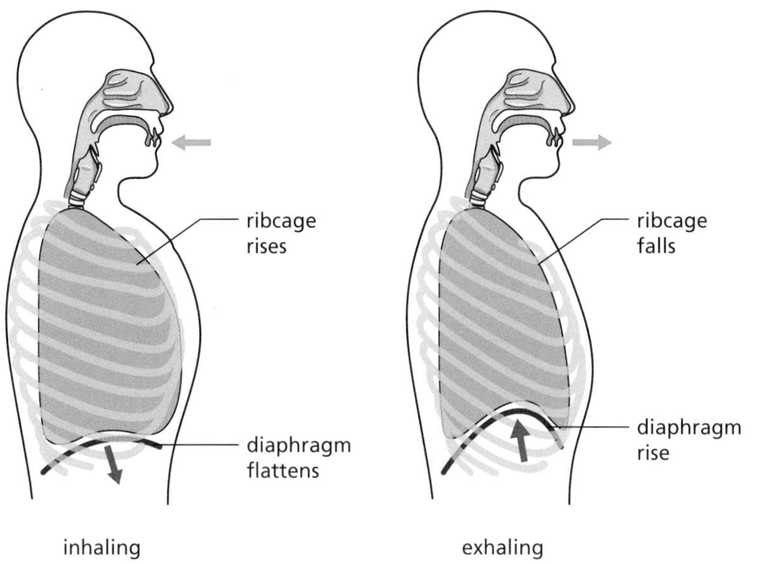

2.17 *Inhaling and exhaling.*

10 Place your hands on the front of your chest and inhale.

a) Describe what you feel.

b) Explain why you feel this.

11 **a)** Describe how diaphragm muscles increase the volume of the chest.

b) What happens to your lungs when your chest volume increases?

12 A student inhales and exhales six times in 30 seconds. What is the breathing rate?

Key terms

breathing rate: the number of times you inhale and exhale in one minute.

exhale: breathing out.

inhale: breathing in.

Activity 2.6: Investigating changes in breathing rate

What effect does exercise have on breathing rate?

A1 Write down a scientific question that you want to answer.

A2 Plan an investigation method to answer your question. Make sure you include the variable you will change, the variable you will measure and any control variables.

A3 State at least *one* way to control risks.

A4 Predict what will happen and explain your prediction.

A5 Present your results neatly.

A6 Compare your prediction with your results, and make a conclusion.

The movement of muscles in your respiratory system is **breathing**. Figure 2.18 shows a model that explains how these muscle movements cause air to move in and out of your lungs.

In the model, pulling the membrane down increases the volume inside the bell jar. This decreases the pressure inside the bell jar (including inside the balloon). Air moves from higher pressure to lower pressure, and so flows into the balloon.

Key term

breathing: movements of muscles in your respiratory system that cause air to move in and out of your lungs.

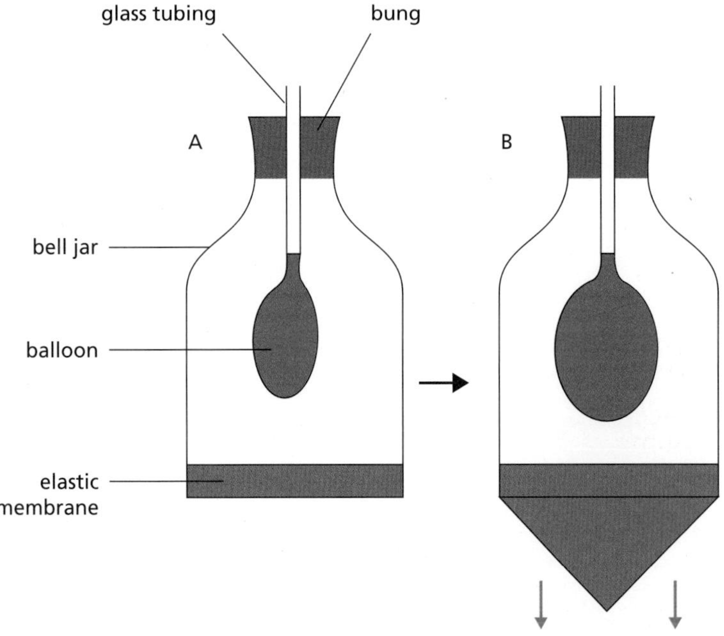

2.18 *A model to explain how breathing causes air to move. The inside of the bell jar represents the inside of the chest.*

Air pollution

Vehicle engines that use petrol and diesel produce many substances that can harm our respiratory and circulatory systems. The World Health Organization thinks that about 7 million people die each year due to air pollution. Many cities are trying to reduce air pollution. In the United Kingdom, people are charged if they want to drive into the centre of London. In India, the authorities have banned all large diesel cars in Delhi. In Brazil, 70% of the people who live in Curitiba use very cheap buses to get around so that the streets are free from cars. Other ideas include banning cars from certain parts of cities, encouraging people to use bicycles and electric vehicles. What are people doing in your country to tackle the problem of air pollution?

13 What is the difference between respiration and breathing?

14 Look at the model in figure 2.18.

- **a)** What does the glass tubing represent?
- **b)** What does the balloon represent?
- **c)** What does the elastic membrane represent?
- **d)** Imagine that you have pulled down the elastic membrane. Explain what happens to the air in the balloon when you release the membrane.

15 Evaluate the strengths and weaknesses of the model in figure 2.18.

Activity 2.7: Investigating lung volume

How can we measure lung volume?

We use the apparatus in the diagram to measure 'lung volume'. This is the maximum volume of air you are able to exhale after a big breath. Adult men have an average lung volume of 4.8 litres. Adult women have an average lung volume of 3.1 litres.

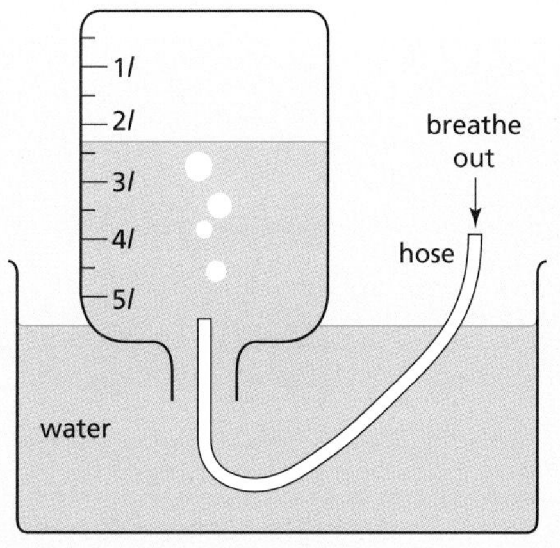

2.19 *Apparatus to measure lung volume.*

A1 Describe how to use the apparatus to measure lung volume.

A2 State at least *one* way to control risks.

A3 Predict how your lung volume compares with the values above. Explain your prediction.

A4 Compare your prediction with your results.

A5 Collect the lung volumes from everyone in your class. Choose some groups for your data and present your data as a tally chart.

A6 Use your tally chart to draw a bar chart.

Key facts:

- ✔ The respiratory system contains many organs, including the lungs, diaphragm and trachea.
- ✔ Breathing is the movement of muscles in the respiratory system.
- ✔ Breathing changes the pressure inside the chest, which causes air to enter or leave the lungs.
- ✔ Gaseous exchange occurs when oxygen diffuses from the alveoli into the blood and carbon dioxide diffuses in the opposite direction.
- ✔ Aerobic respiration can be shown using a word equation:

  oxygen + glucose → carbon dioxide + water.

Check your skills progress:

I can plan investigations to test ideas.

I can make and explain predictions.

I can present results using tables.

I can divide data into smaller groups of equal size.

I can select and draw the correct type of bar chart to show my continuous and discontinuous data.

Smoking and health

Learning outcomes

- To describe how your body keeps your lungs clean
- To explain why your lungs need to be kept clean
- To describe some of the problems caused by smoking

Starting point

You should know that...	You should be able to...
Your body has specialised cells to do certain functions	Research and present data using secondary sources
Gaseous exchange happens in the alveoli in the lungs	Use bar charts and line graphs to present data

Dust, viruses and bacteria may all damage your lungs. Your respiratory system has ways to stop them getting into your lungs.

Nose hairs filter air to remove larger particles. Smaller particles stick to a thick liquid **mucus**, produced by cells in the tubes of your respiratory system. These smaller particles include microorganisms, which can grow and reproduce in the mucus. However, **ciliated epithelial cells** use cilia to sweep the mucus up to your throat. You then swallow it and acid in your stomach kills the microorganisms.

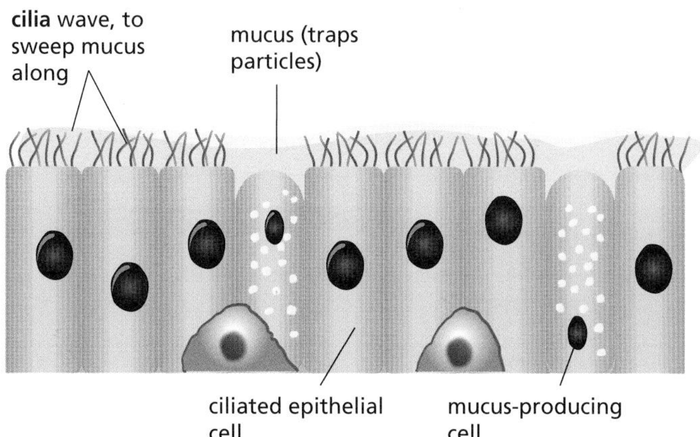

2.20 *Ciliated epithelial cells help keep the lungs clean.*

1 Why is it important to stop particles getting into your lungs?

2 **a)** How are ciliated epithelial cells adapted for their function?

b) What other specialised cells are needed to keep the lungs clean?

Key terms

cilia: waving strands that stick out of some cells.

ciliated epithelial cell: specialised cell with waving cilia to sweep mucus along.

mucus: sticky liquid that traps particles.

Smoking

Many people smoke tobacco (mainly in cigarettes). The heat and substances in tobacco smoke damage cilia, and paralyse them. They no longer wave and so cannot sweep up mucus. Smoking for many years destroys the cilia. Long-term smokers must cough up the mucus to get rid of it. This is a 'smoker's cough'.

3 In the lungs of a smoker, why do the cilia stop working?

4 Explain why long-term smokers are more likely to get lung infections.

5 Coughing, and the substances in smoke, can destroy alveoli. This causes people to feel breathless because of poor gaseous exchange. Explain why destruction of alveoli causes poor gaseous exchange.

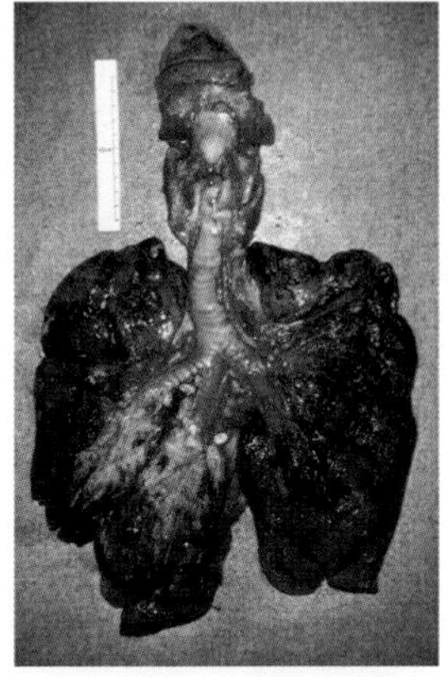

2.21 *Lungs showing the effects of tobacco smoke.*

A **drug** is a substance that affects how your body works. Tobacco smoke contains a drug called **nicotine**, which affects your brain and increases heart rate. This drug is **addictive**, which means that people feel that they need to have it. Over time, nicotine damages arteries and makes plaques more likely to form.

6 **a)** Describe one instant effect that nicotine has on your body.

b) Explain how nicotine damages the circulatory system with time.

Tobacco smoke contains a black, sticky mixture of substances called **tar**. This may cause **cancer**, in which a tissue makes new cells in an uncontrollable way. The new cells may form a lump called a **tumour**.

7 What is a cancer tumour?

8 Look at figure 2.21. Lung tissue is pink. Why is the tissue not pink in this lung?

Scientists ask questions about why things happen. They then create ideas that might explain why things happen. They test those ideas using experiments or observations.

Between 1900 and 1950 in the United Kingdom, there was a big increase in lung cancer. A doctor, Richard Doll, created three possible ideas to explain this:

Key terms

addictive: substance that makes people feel that they must have it.

cancer: when cells in a tissue start to make many copies of themselves very quickly.

drug: substance that affects the way your body works.

nicotine: addictive drug in tobacco smoke.

tar: sticky black liquid found in cigarette smoke.

tumour: a lump of cancer cells.

- increase in road building (and the tar used)
- increase in traffic (and exhaust fumes)
- increase in smoking.

He made many observations. He found that the more people smoked, the more lung cancer deaths there were. The two variables changed together. This is a **correlation**.

We use a **scatter graph** to look for a correlation between two variables. These variables must both be measured as numbers. You plot the variables on a graph and look for a pattern. For example, you might look for a line of points. You might be able to draw a **line of best fit** through the points.

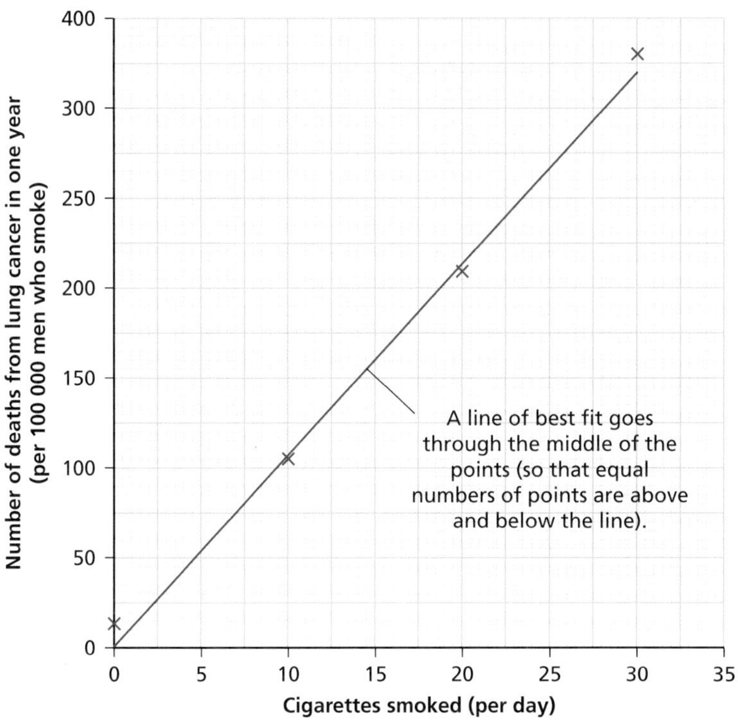

2.22 *The line on this scatter graph shows a correlation between the number of cigarettes smoked and the number of deaths from lung cancer.*

9 Look at the graph.

a) What type of graph is this?

b) Describe the correlation shown.

Key terms

correlation: relationship (link) between variables where one increases or decreases as the other increases.

line of best fit: straight or curved line drawn through the middle of a set of points to show the pattern of data points.

scatter graph: graph of two variables, both measured in numbers.

Smoking tobacco

A scientific study found that between 1990 and 2015, smoking caused one in every 10 deaths in the world. It also found that the number of smokers in the world is increasing, because the world population is increasing.

The percentage of people who smoke in some countries has fallen (such as Brazil, Australia and Nigeria). In others, there was no change (such as Bangladesh, Indonesia and the Philippines). The percentage of smokers in some groups has increased (such as women in Russia).

Smoking creates jobs, such as tobacco farming. Governments also get money from taxes on tobacco. But smoking causes diseases and many countries try to reduce the number of smokers. They put warnings on cigarettes and raise the taxes on them.

Activity 2.8: Investigating smoking

What are the 'pros' and 'cons' of smoking?

A1 Use this and other books and/or the internet to:

- find some good points about smoking
- find some bad points about smoking
- discover what your country is doing about smoking.

A2 Write a report of three or four paragraphs. End your report by saying what you think about smoking. Your teacher may ask you to use your ideas in a debate.

10 Scientists use the apparatus in figure 2.23 to investigate cigarette smoke.

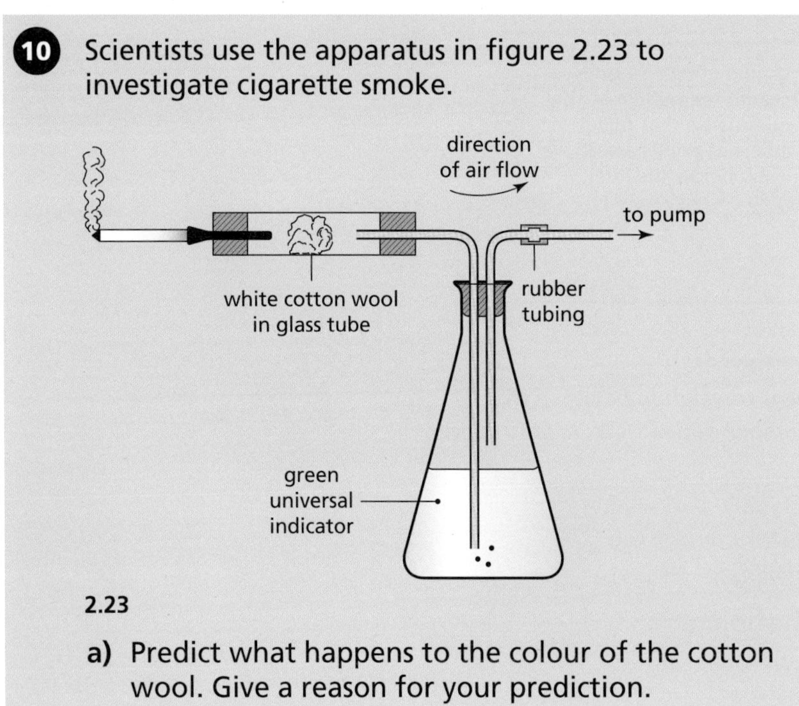

2.23

a) Predict what happens to the colour of the cotton wool. Give a reason for your prediction.

b) A scientist puts a thermometer in the glass tube. Will the temperature reading be higher, lower or the same as outside the apparatus?

c) The universal indicator turns orange. What does this tell you about cigarette smoke?

11 Tar coats the inside of the lungs. Explain what effect this has on gaseous exchange.

Key facts:

- ✔ Specialised cells keep your lungs clean (such as cells that produce mucus and ciliated epithelial cells).
- ✔ Tobacco smoke paralyses cilia, and contains an addictive drug called nicotine.
- ✔ Tar in tobacco smoke causes cancer.

Check your skills progress:

I can look for patterns and correlations in data.

I can interpret scatter graphs (and lines of best fit) to look for correlations.

I can describe how scientists answer their questions by creating ideas that they test.

Transport of water and mineral salts in plants

Learning outcomes

- To explain what a plant needs water and mineral salts for
- To describe how water enters a plant
- To describe the route that water takes through a plant

Starting point

You should know that...	You should be able to...
Plants have organs, such as roots, stems and leaves	Think up scientific questions to ask
Plants need water	Plan investigations to test ideas
Plants have specialised cells, such as root hair cells and guard cells (that have stomata between them)	Make predictions and conclusions

Roots hold a plant in the ground. Roots also absorb water and mineral salts.

Mineral salts are substances that plants need in very small amounts. Plants use them to make some substances in their biomass. For example, nitrates are mineral salts that plants need to make proteins. Like us, plants need proteins for growth and repair. Other mineral salts contain magnesium, which a plant needs to make chlorophyll.

A plant needs water to make its own food, using a process called photosynthesis. It also needs water to give cells their shapes. Without enough water, plant cells start to collapse and the plant **wilts** (becomes floppy).

Key term

wilting: when a plant becomes floppy due to lack of water.

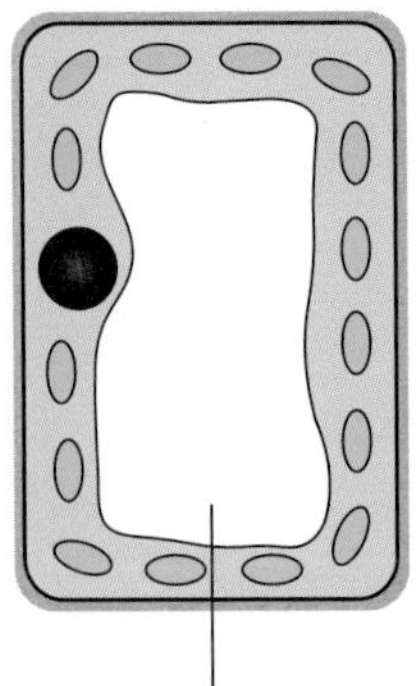

When filled with enough water, the vacuole pushes outwards on the cell. This keeps its shape.

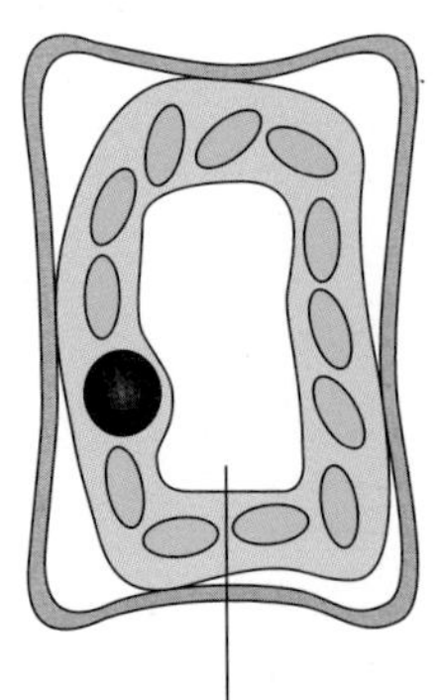

Without enough water, the vacuole shrinks. There is no longer enough force pushing outwards. The cell wall starts to bend and the cell loses its shape.

2.24 *Plant cells need water to keep their shapes.*

1 Describe *two* functions of roots.

2 a) Explain why a plant that lacks nitrates does not grow very well.

b) Explain why a plant that lacks magnesium has yellow leaves.

c) Explain why a plant wilts when it lacks water.

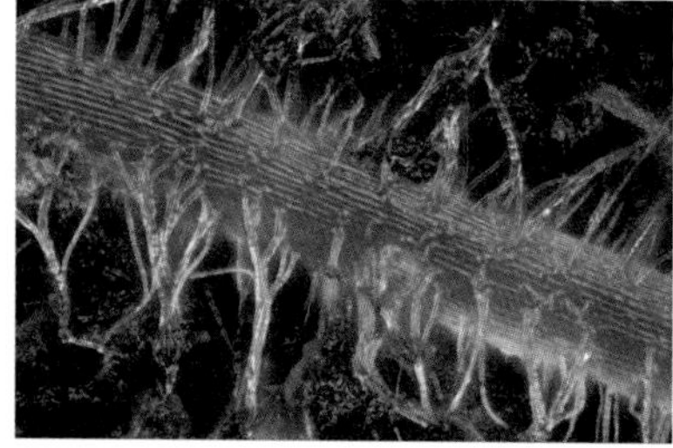

2.25 *Root hair tissue magnified x25 times.*

Absorbing water

Roots use specialised **root hair cells** to **absorb** (take in) water. These cells have bits sticking out them that look a bit like hairs. A 'root hair' gives a cell a lot of surface area, which helps it absorb water quickly. These cells also absorb mineral salts.

Key terms

absorb: to take in or soak up.

root hair cell: plant cell found in roots that is adapted for taking in water quickly.

3 a) What do root hairs absorb?

b) Explain how 'root hairs' help cells to absorb substances quickly.

4 What apparatus has magnified the cells in figure 2.25?

5 Make a labelled drawing of a root hair cell, showing all its parts.

Transporting water

The stem is a plant organ that supports a plant and holds its leaves in place. It also transports substances.

A plant transports water and minerals salts in specialised xylem cells. These cells form chains and then die, to form hollow tubes. The tubes run continuously from the roots up to the leaves. They have thick cell walls, containing lignin. This substance makes the walls very strong and stops them collapsing.

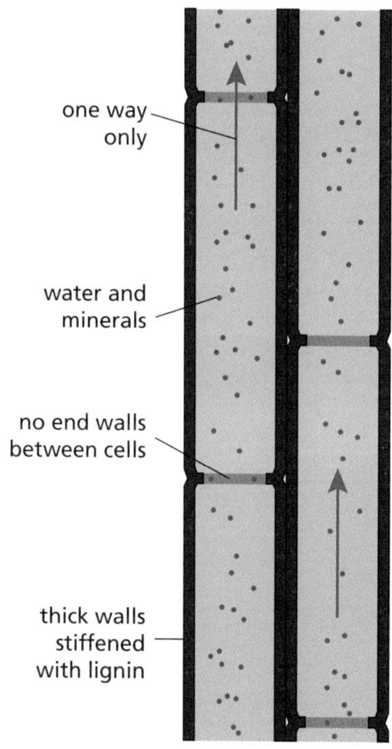

2.26 *Xylem cells die and form hollow tubes.*

6 How are xylem cells adapted to their function?

Transpiration

Water travels up to the leaves of a plant, where it enters cells and evaporates into the air spaces. Water vapour exits the leaves through stomata. As water is lost, it pulls more water up through the xylem and into the leaves. The word **transpiration** describes the route of water into the roots, up the stem and out through the leaves.

Key term

transpiration: the flow of water into a plant's root, up its stem and out of its leaves.

7 How is water lost from a plant?

8 What happens in transpiration?

Activity 2.9: Investigating water transport in stems

What variables affect how far water moves in a stem in a certain time?

Water rises up some plant stems when placed in water. Colouring the water helps you to see this. You can find the level that the water reaches by cutting the stem.

A1 Think of a variable that might affect how quickly water flows up a stem. Write down a scientific question that you want to answer.

A2 Plan an investigation method to answer your question. Make sure you include the variable you will change, the variable you will measure and any control variables.

A3 Predict what will happen and explain your prediction.

A4 Present your results neatly.

A5 Compare your prediction with your results, and make a conclusion.

Phloem

Phloem cells form long chains of living cells. The cells transport soluble sugars up and down a plant.

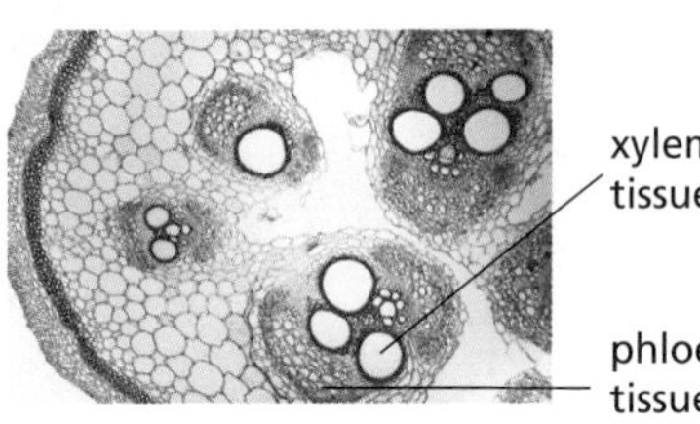

2.27 *Xylem and phloem tissue in a stem.*

9 What substances do these tissues transport?

a) phloem

b) xylem

10 List the organs in a plant's 'water transport system'.

11 How are xylem and phloem tissues different?

Key term

phloem cell: plant cell that is adapted to form living tubes to transport sugars and other substances.

Key facts:

- ✔ Plants need water to make their own food (by photosynthesis) and for cells to keep their shapes.
- ✔ A plant needs mineral salts to make substances (such as proteins).
- ✔ Water and mineral salts are absorbed by a plant using root hair cells.
- ✔ Water and mineral salts are transported in tubes formed by dead xylem cells.
- ✔ Water is lost from a plant through its stomata.

Check your skills progress:

I can develop scientific questions to investigate.

I can plan investigations, to answer scientific questions by choosing which variables to change and to measure.

I can plan investigations, choosing which variables to control.

End of chapter review

Quick questions

1. Your heart, blood and blood vessels are all part of your:

   a respiratory system b circulatory system
   c excretory system d nervous system [1]

2. Valves are found in:

   a veins b arteries c capillaries d veins and arteries [1]

3. Urea is mainly excreted by your:

   a liver b lungs c heart d kidneys [1]

4. The passage of water through a plant, in at the roots and out at the leaves, is called:

   a translocation b transport c transpiration d transcription [1]

5. Another name for the trachea is:

   a bronchus b windpipe c oesophagus d ventricle [1]

6. (a) In which direction do arteries carry blood? [1]

   (b) Describe *one* way in which arteries are adapted to their function. [1]

7. Give a reason why a plant needs:

   (a) water [1]

   (b) mineral salts. [1]

8. Write out the word equation for aerobic respiration. [2]

9. Name the addictive drug in tobacco smoke. [1]

10. Blood from most of your body travels to your heart in a large blood vessel called the vena cava. Which heart chamber does this blood enter? [1]

11. Describe the functions of:

    (a) red blood cells [1]

    (b) white blood cells [1]

    (c) platelets. [1]

Connect your understanding

12. **(a)** Which parts of your respiratory system move to make your chest volume increase? **[2]**

(b) Explain why air flows into your lungs when your chest volume increases. **[2]**

13. A student breathes in and out 7 times in 30 seconds. Calculate the breathing rate. **[1]**

14. When Noor is 40, a plaque forms in one of his large arteries. It grows as he gets older. When he is 65, the plaque splits and a blood clot forms on it. The clot breaks off the plaque.

(a) Suggest a reason why the plaque formed. **[1]**

(b) Explain why the plaque causes Noor's blood pressure to rise. **[2]**

(c) Describe *one* problem that high blood pressure can have. **[1]**

(d) Noor has a heart attack when he is 65. Explain how this happens. **[3]**

15. The drawing shows a part of a lung where gaseous exchange occurs.

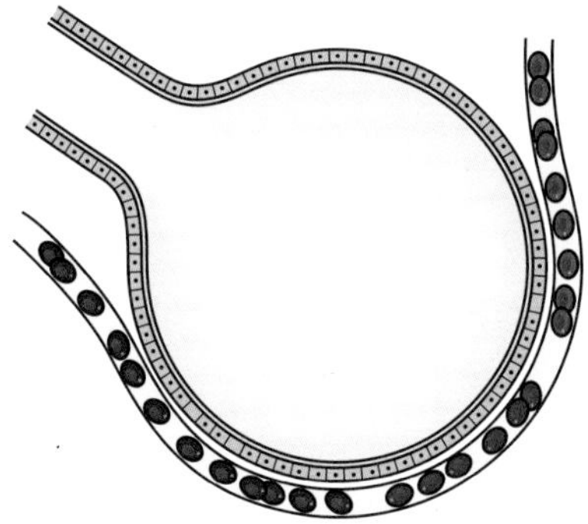

2.28

(a) What is this part called? **[1]**

(b) Give the names of the gases that are exchanged. **[2]**

(c) What process causes the gases to move? **[1]**

(d) Explain how the structures shown in the diagram are adapted to help the gaseous exchange. **[2]**

(e) Explain why the red blood cells are not all the same colour in the diagram. **[1]**

(f) Explain *one* difference between the plasma at the top of the diagram and at the bottom of the diagram. **[2]**

16. In an investigation, some athletes ran at different speeds for 5 minutes. A scientist then measured their pulse rates. The graph shows the results.

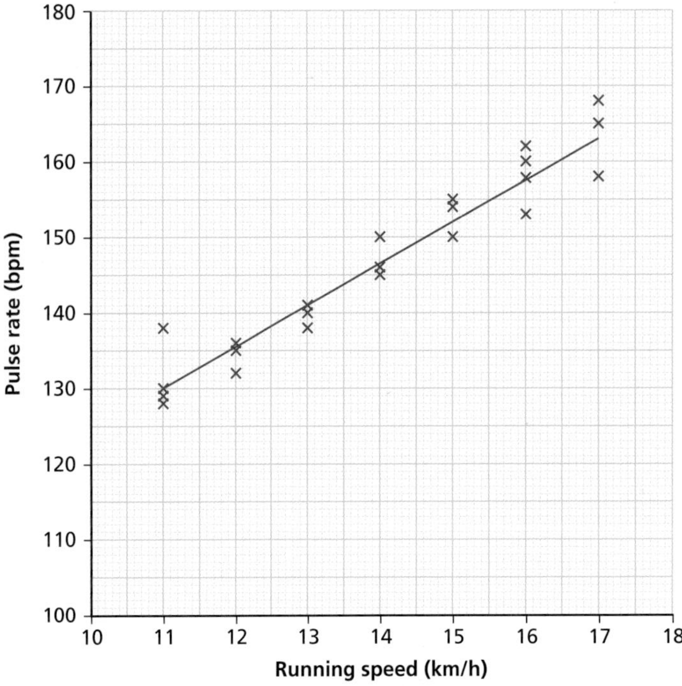

2.29

(a) What is a pulse? **[1]**

(b) Use the line of best fit on the graph to give the pulse rate for an athlete running at 12.8 km/h. **[1]**

(c) What correlation is shown in the graph? **[1]**

(d) Explain the correlation. **[3]**

17. **(a)** Draw a labelled diagram to explain how specialised cells in the tubes of your lungs keep them clean. **[2]**

(b) Explain what happens if these cells stop working. **[2]**

Challenge questions

18. Carbon monoxide is a poisonous gas, which can kill. It sticks to haemoglobin. This stops other substances sticking to haemoglobin. One of the early symptoms of carbon monoxide poisoning is an increase in breathing rate. Explain this symptom. **[3]**

19. In an experiment, a scientist gave people different amounts of nicotine. The scientist then measured the speed of the blood in skin capillaries. After the experiment, the speed of the blood quickly returned to normal.

Mass of nicotine taken into the body (mg)	**Average speed of blood in the capillaries (mm^3/sec)**
0	0.000 047
0.50	0.000 036
0.98	0.000 034
1.90	0.000 027

(a) What is the correlation shown in this data? **[1]**

(b) Suggest what nicotine does to capillaries to cause this effect. **[1]**

20. An increase in temperature increases the speed of diffusion. Explain why plants are more likely to wilt in very hot weather. **[3]**

3

Chapter 3

Reproduction and growth

What's it all about?

Reproduction is one of life's amazing wonders. Two cells join together and a new life forms! The two cells are an egg cell from a female and a sperm cell from a male. A male releases many sperm cells, but a female usually releases only one egg cell at a time. So, the race is on to see which sperm cell will get to the egg cell and join with it first!

You will learn about:

- The human reproductive system and how the female reproductive system is adapted to support the growth of new life
- The changes that happen to your body and emotions during adolescence
- The menstrual cycle and fertilisation of a human egg by a sperm
- The effects that diet, drugs and disease can have on your body during growth and development

You will build your skills in:

- Using data from other sources
- Using graphs to explain scientific ideas

Reproductive systems

Learning outcomes

- To describe the parts and functions of the human reproductive system
- To explain how human gametes are adapted for their functions
- To describe what happens during fertilisation and how a fertilised egg cell develops into an embryo

Starting point

You should know that...	You should be able to...
Reproduction is one of the seven life processes	Collect and use information from secondary sources
Males and females are involved in human reproduction	

Human reproduction

Like any other organism, humans have to reproduce to make new individuals. Humans reproduce using **sexual reproduction**, which involves a male sex cell and a female sex cell. The nuclei of these sex cells (or **gametes**) join together in a process called **fertilisation**. In humans, this happens inside the female's body so we call it **internal fertilisation**.

Fertilisation produces a fertilised egg cell or **zygote**. This then grows and develops inside the female, to form an **embryo** that then grows into a baby.

What happens during sexual reproduction?

Gametes

There are two types of human gamete. Egg cells are the gametes made by females. Sperm cells are the gametes made by males.

Key terms

embryo: small ball of cells that develops from a fertilised egg cell. It becomes attached to the uterus lining and develops into a foetus.

fertilisation: when an egg cell nucleus and a sperm cell nucleus fuse (join) and form a fertilised egg cell.

gamete: sex cell (egg cell in a female and sperm cell in a male).

internal fertilisation: when the nuclei of the egg and sperm join together inside the female animal.

sexual reproduction: the type of reproduction involving male and female gametes coming together.

zygote: fertilised egg cell.

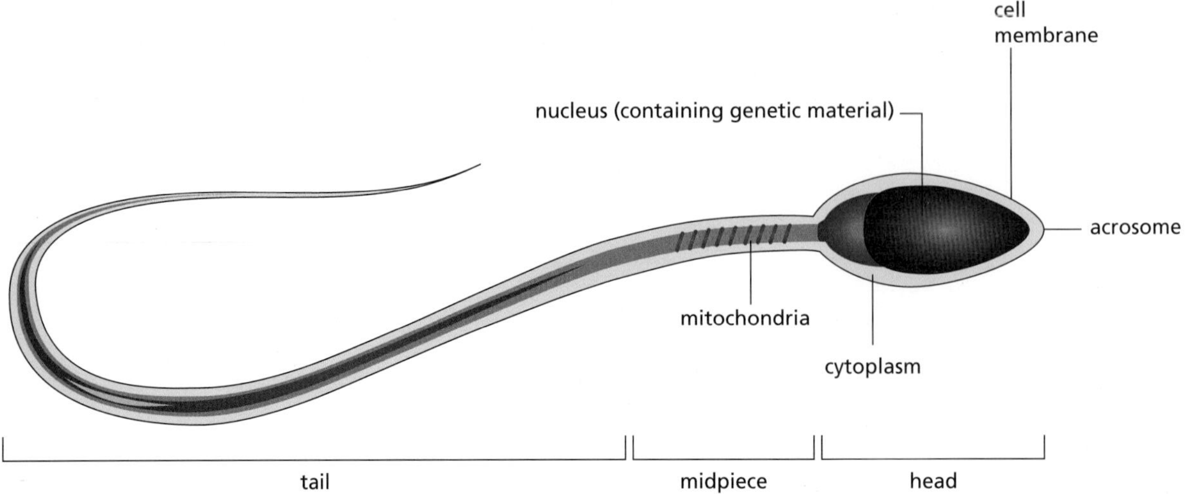

3.1 *A human sperm cell.*

Sperm cells are **microscopic**. They are only about 0.05 mm long. From puberty onwards, the testes of males make sperm cells continuously (figure 3.1). Sperm are specialised cells and are adapted to carry out their function in the following ways.

- A sperm cell has a tail to allow it to swim towards the egg cell.
- The sperm cell nucleus contains half the instructions for making a new human.
- A sperm cell has many **mitochondria** (arranged in a spiral shape around the top of the tail), which release energy using aerobic respiration. The sperm cell can use this energy to move its tail and swim towards the egg cell.
- The tip of the head of the sperm cell, known as the acrosome, makes a special substance that can help it to break into the egg cell.

Key terms

microscopic: something so small you can only see it using a microscope.

mitochondria: part of a cell where aerobic respiration happens to release energy. The singular form is 'mitochondrion'.

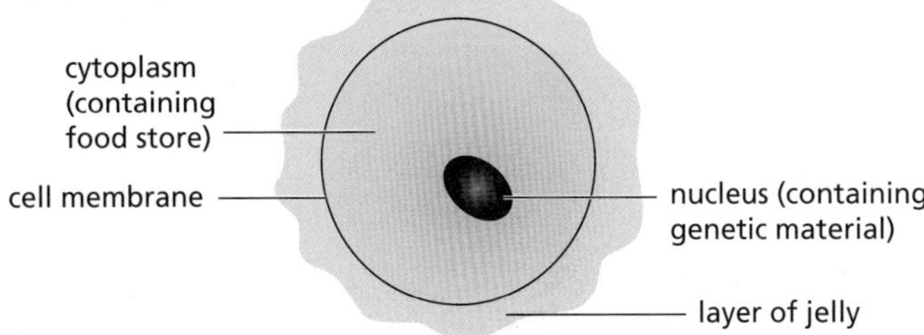

3.2 *A human egg cell.*

Egg cells are approximately 30 times larger than sperm cells. They are large enough for you to see them without a microscope. Females are born with all the eggs that they will ever have. The egg cells are stored inside the **ovaries**

Key term

ovary: female reproductive organ where eggs are made, stored and matured. Females have two ovaries.

(figure 3.4). Eggs are specialised cells and are adapted to carry out their function in the following ways.

- Egg cells have food reserves in the cytoplasm. The cytoplasm is where the mitochondria are found in the egg cell. These food reserves can be used by the zygote as it develops. The zygote develops into a **foetus**.
- The egg cell nucleus contains half the instructions to make a human.
- There is a layer of jelly surrounding the egg cell to make sure that only one sperm cell can enter.

Key term

foetus: baby developing in a woman's uterus from about 10 weeks of development, when it starts to resemble a baby.

2 What are the *two* human gametes?

3 Give *three* adaptations of a sperm cell that make it good at its job.

4 Explain why it is important for the egg cell to have only half of the instructions to make a new human.

Male reproductive system

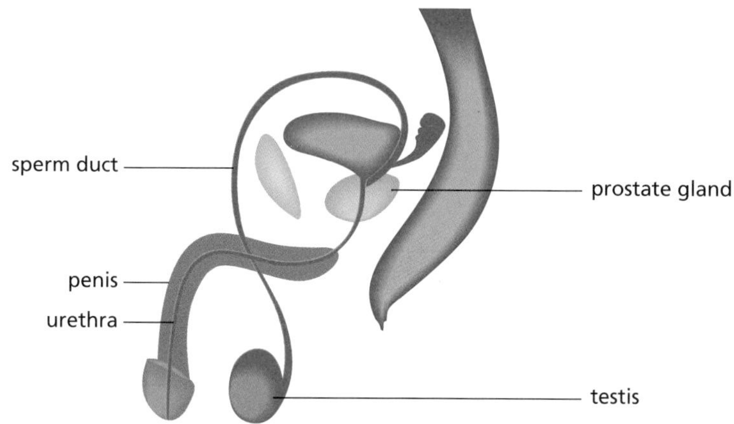

3.3 *The male reproductive system.*

5 What is the function of the testes?

6 What happens in the ovaries?

Key terms

penis: male reproductive organ used to transfer sperm to female cervix during intercourse.

prostate gland: gland that surrounds the bottom of a male's bladder. It produces some of the liquid that makes up **semen**.

semen: the liquid containing sperm.

sperm duct: the tube that carries sperm from the testis to the urethra.

testis: (plural testes) male organ where sperm are made. Males have two testes.

Female reproductive system

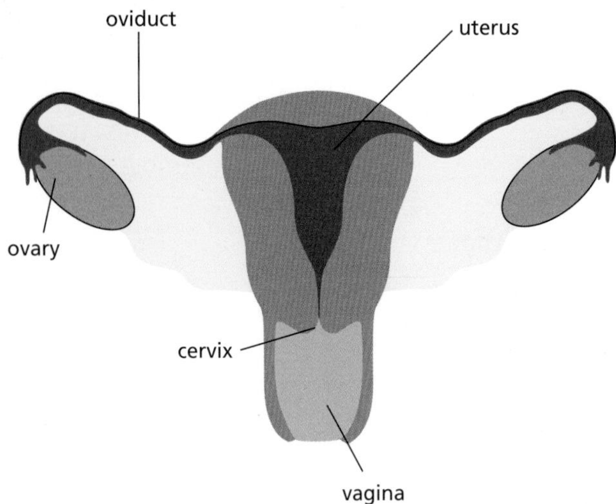

3.4 *The female reproductive system. (The bladder is not shown.)*

Key terms

cervix: the neck of the uterus.

oviduct: the tube which connects the ovary to the uterus. Females have two oviducts, one from each ovary.

uterus: female reproductive organ where the baby grows when a woman is pregnant.

vagina: the tube joining the uterus to the outside of the female body.

Fertilisation

During sexual intercourse, sperm cells are transferred from the **penis** of the male into the **cervix** of the female. From here, the sperm cells swim through the **uterus** and towards the **oviduct** (figure 3.4). If an egg cell has recently been released from an ovary, sperm cells may meet it in the oviduct.

Sperm cells release a special substance to help them break through the jelly layer and cell membrane of the egg. Only one sperm cell can break through the jelly coat of the egg cell. Once the sperm cell has broken through the layers, the nuclei of the sperm cell and egg cell fuse (join) together. This is fertilisation.

The sperm cell nucleus and egg cell nucleus each have only half the information needed to make a new human. When they fuse, the new cell (fertilised egg cell) has a full set of information needed. The fertilised egg cell begins to divide to form a ball of cells, then an embryo, and eventually a foetus.

7. Describe the path a sperm cell takes once it enters the female body.
8. What happens during fertilisation?
9. Where does fertilisation usually happen?
10. Why is it important that the nucleus from the egg cell and the sperm 'fuse' during fertilisation?

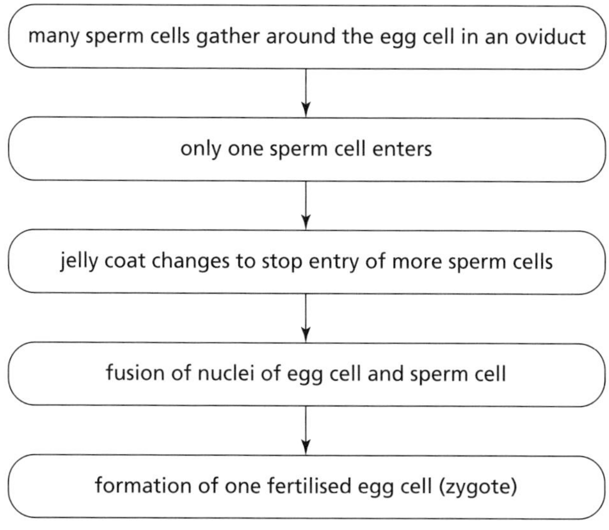

3.5 *Fertilisation and the stages leading up to it.*

Activity 3.1: Researching fertilisation

Research the differences between internal fertilisation and external fertilisation.

A1 Discover the names of three animals that use internal fertilisation and three that use external fertilisation.

A2 What are the main differences between internal fertilisation and external fertilisation?

The man who 'discovered' sexual reproduction in animals

An Italian called Spallanzani was the first person to show that sexual reproduction needs both a sperm cell and an egg cell. He was also the first to perform *in vitro* (in glass) fertilisation using frogs. (He mixed frog sperm and egg cells in a dish. Some of them fused to make fertilised egg cells!)

Key facts:

- ✔ Male gametes are sperm cells and female gametes are egg cells.
- ✔ Sperm cells and egg cells have many adaptations for their functions.
- ✔ When a sperm cell nucleus and an egg cell nucleus fuse, a fertilised egg is made. This develops into an embryo, which develops into a foetus, which is born and called the baby.

Check your skills progress:

I can use information from diagrams to help me explain scientific ideas.

Puberty

Learning outcomes

- To describe the stages of human growth and development
- To describe what happens during puberty and adolescence
- To describe what happens in the menstrual cycle

Starting point

You should know that...	You should be able to...
Humans go through different stages of life	Interpret information from a diagram or chart
	Use tables and line graphs

The human life stages

The first stage of human life happens inside a woman's uterus as her baby develops. When the baby is born it is an **infant**. Humans are infants for the first year after birth and then they enter the life stage we call childhood.

At about 11 years old humans become adolescents – the age varies for different people. During **adolescence**, humans experience emotional and physical changes, and **puberty** occurs. Puberty is when physical changes occur. At about age 18 adolescents become adults. Adulthood is the longest life stage, and lasts until old age.

A human grows from the zygote stage until adulthood. Figure 3.6 shows how the height of a human changes from birth. Young babies are monitored to ensure that their growth is somewhere near the spread of heights on the graph, and that they continue to grow. If either of these are not happening, then medical intervention may be needed.

1. What is the average height of baby boys when they are born?
2. What are the highest and lowest heights of 10-year-olds shown by this graph?
3. On average, at what age do girls appear to stop getting taller?
4. Describe how the heights of girls and boys change between the ages of 10 and 18.

Key terms

adolescence: the life stage in humans that usually happens between the ages of 11 and 18. During this stage, people go through many emotional and physical changes.

infant: the life stage which lasts for the first year after birth.

puberty: the physical changes that happen to the body during adolescence.

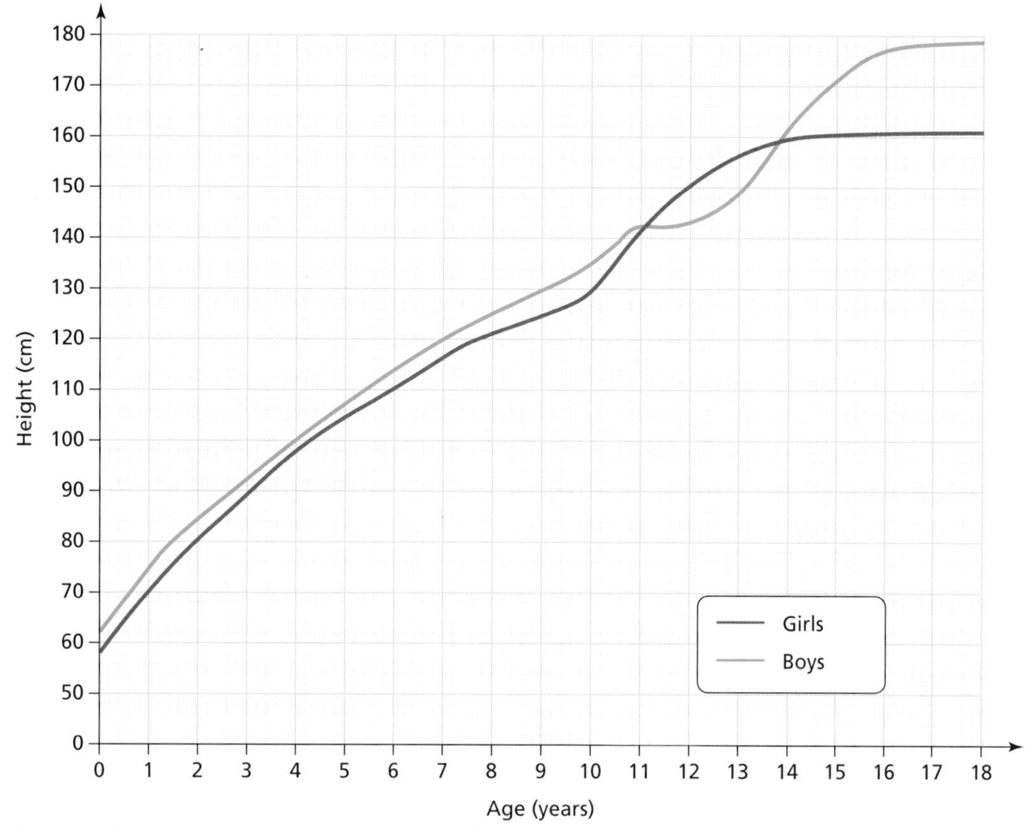

3.6 *Graph showing the average heights of girls and boys from birth to age 18. The steeper the line, the faster the growth.*

Activity 3.2: Investigating hamster growth

How quickly does a hamster grow?

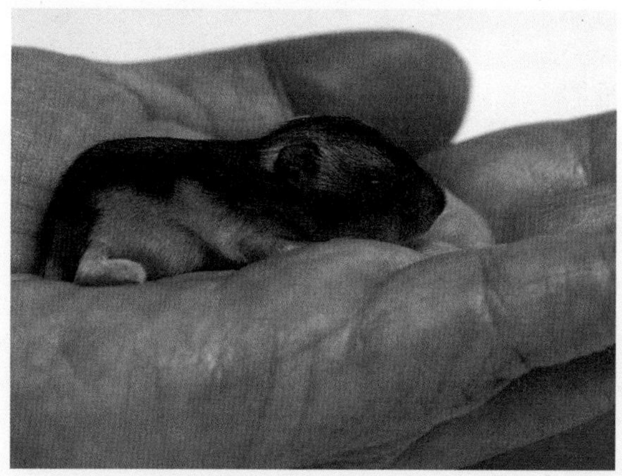

3.7 *A baby hamster.*

Kala and Sabina have a baby hamster. They decided to measure his mass every week to see how quickly he was growing. Their results are shown in the table.

Week number	1	2	3	4	5	6	7	8
Mass of hamster (g)	50	60	105	120	140	150	155	155

A1 Name *two* things that the girls should have controlled (kept the same) each time they measured the mass of the hamster.

A2 Plot a line graph to show the growth of the hamster.

A3 Looking at your graph, between which two weeks did the hamster grow the fastest?

A4 Would all hamsters follow the same growth pattern as this? Explain your answer.

Puberty

When you are aged 11–18, you are in adolescence. This time can be very difficult because there are many emotional changes as well as physical changes. Chemical substances called **hormones** cause many of these changes.

Hormones are made in different parts of your body and travel in your bloodstream. They cause certain parts of your body to do certain things. **Testosterone** is a hormone that is especially important in boys, and an important hormone in girls is **oestrogen**. These hormones are important as they ensure that **secondary sexual characteristics** develop.

Puberty is when physical changes happen during adolescence. During these changes your body develops secondary sexual characteristics. The age when puberty begins varies for different people. For some, it can begin as young as seven, whereas others may be older than 18. Puberty, and when it starts, is not something to worry about.

What cause the physical and emotional changes that occur during adolescence?

Key terms

hormone: chemical released into the bloodstream which has an effect on certain parts of your body.

oestrogen: hormone that triggers many of the physical changes in girls during puberty.

secondary sexual characteristics: the physical changes that happen to a person's body during puberty.

testosterone: hormone that triggers many of the physical changes in boys during puberty.

Secondary sexual characteristics

When puberty begins, people begin to notice that their bodies develop secondary sexual characteristics. These physical changes vary between boys and girls. Table 3.1 summarises the main changes.

Physical changes in females during puberty	Physical changes in males during puberty
Growth spurt	Growth spurt
Body hair increases under arms and in pubic region	Hair growth increases, particularly on face, under arms and in pubic region
Body odour increases	Body odour increases
Breasts develop	Voice deepens
Hips widen	Chest and shoulders broaden
Menstrual cycle begins	Penis and testes increase in size
	Ability to ejaculate

Table 3.1 *Secondary sexual characteristics in females and males.*

6 Give *three* physical changes that girls go through during puberty.

7 Give *three* different physical changes that boys go through during puberty.

Extreme facial hair!

This man is showing off his very long facial hair in a moustache competition at Pushkar fair in India. Every year pilgrims and traders gather in the town of Pushkar to trade camels and livestock as well as to take part in competitions such as the 'longest moustache'!

3.8

Emotional changes

Hormones cause emotional changes as well as physical changes. During puberty, people often experience mood swings, low self-esteem and uncertainty. All of these feelings are normal. If you talk to friends, you will probably find that many of them have experienced the same things.

The menstrual cycle

One change that happens in girls during puberty is that the **menstrual cycle** begins. This means the ovaries begin to release egg cells regularly – usually one ovary releases an egg cell about every 24–35 days. The menstrual cycle continues then until the woman reaches **menopause** (around age 45–55).

The length of the menstrual cycle is different for different women, but usually lasts about 24–35 days. There are different phases during the menstrual cycle (figure 3.9).

Key terms

menopause: the time in a woman's life when the menstrual cycle stops, normally between ages 45 and 55.

menstrual cycle: cycle of changes that happens in females after puberty. During each cycle an egg cell is released from an ovary and (if it is not fertilised) menstruation happens. Each cycle lasts between about 24 and 35 days.

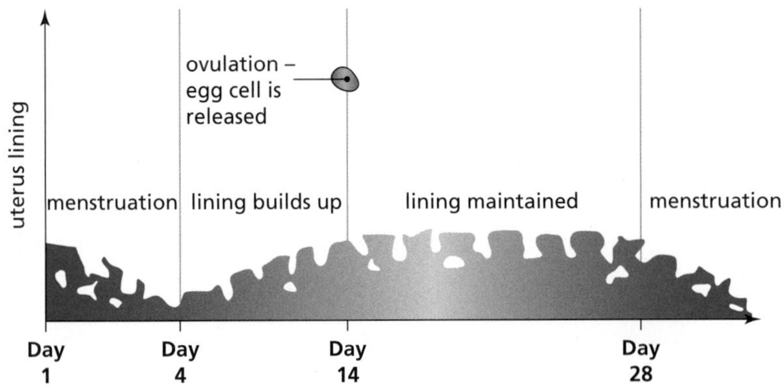

3.9 *What happens during a menstrual cycle that is 28 days long.*

Day 1–4: **Menstruation** – the uterus lining disintegrates and is lost from the woman's body. This is called a **period**.

Day 5–14: The uterus lining builds up again. If an egg cell is fertilised, the lining will be ready to support the developing foetus.

Day 14: **Ovulation** – one of the ovaries releases an egg cell, and it moves down the oviduct.

Day 14–28: The uterus lining stays thick, in case the released egg cell is fertilised.

Day 28/Day 1: The cycle begins again. If the egg cell was not fertilised by a sperm cell, the uterus lining is not needed and so it starts to die and is lost from the woman's body.

If an egg is fertilised, it may attach to the uterus lining. If this happens, the woman becomes **pregnant**. During pregnancy the uterus lining stays thick because it needs to support the developing foetus.

8 **a)** What is menstruation?

b) When in the menstrual cycle does menstruation happen?

9 In a 28-day cycle, on which day is an egg released from the ovary?

10 Why does the uterus lining build up?

11 If a woman has a 28-day cycle and begins menstruating on 3 April, when should she expect to begin her next menstruation period?

Key terms

menstruation: the time in the menstrual cycle when the uterus lining breaks down and is lost. It is called a period.

ovulation: the release of an egg cell from one of the ovaries.

period: stage in the menstrual cycle when the lining of the uterus is lost from the body.

pregnant: a woman becomes pregnant if a fertilised egg implants in her uterus.

Key facts:

- ✔ Human life stages are infancy, childhood, adolescence, adulthood and old age.
- ✔ Puberty is the time, during adolescence, when physical changes occur in the body.
- ✔ During adolescence, people go through many physical and emotional changes.
- ✔ During puberty males and females develop secondary sexual characteristics.
- ✔ The female menstrual cycle begins during puberty and includes ovulation and menstruation.

Check your skills progress:

I can collect information from a diagram or chart and use it to explain a scientific concept.

I can draw a graph as well as interpret and explain what it is showing.

Foetal growth and development

Learning outcomes

- To describe the development of the foetus
- To explain how the foetus is protected and nourished during the gestation period
- To explain how a baby is born

Starting point

You should know that...	You should be able to...
Sperm cells and egg cells are specialised animal cells	Collect and use information from secondary sources
A developing baby is carried inside the mother	Plot a graph

A fertilised egg cell becomes an embryo

Figure 3.10 shows an egg cell being released from the ovary and beginning its journey along the oviduct. If the egg cell meets a sperm cell and fertilisation occurs, the fertilised egg cell (zygote) starts to divide 24 hours later. This cell division continues and very soon a small ball of cells forms, called an embryo.

After about six days, the embryo completes its journey down the oviduct and reaches the uterus. By now, the embryo is made of more than 100 cells! In the uterus, the embryo attaches to the uterus lining. This is called **implantation**.

Key term

implantation: the developing embryo attaches to the uterus lining.

1. What is an embryo?
2. What happens to the embryo when it reaches the uterus?

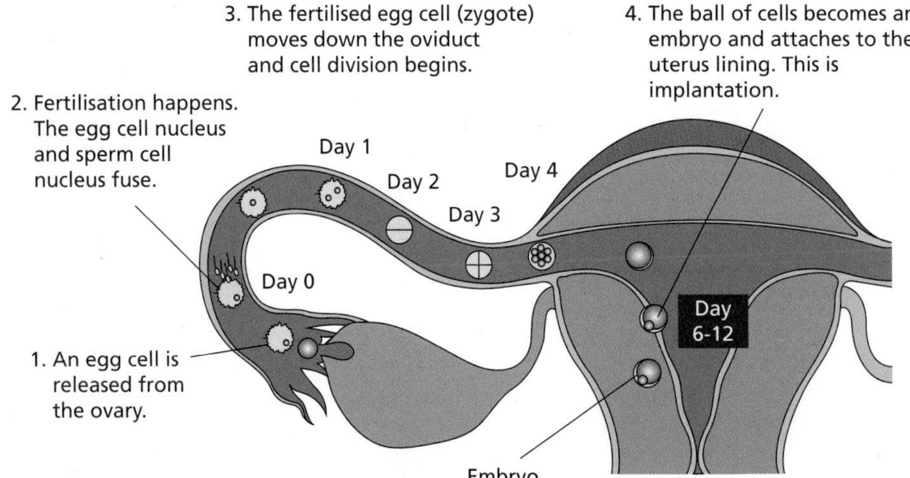
3. The fertilised egg cell (zygote) moves down the oviduct and cell division begins.
4. The ball of cells becomes an embryo and attaches to the uterus lining. This is implantation.
2. Fertilisation happens. The egg cell nucleus and sperm cell nucleus fuse.
Day 1
Day 2
Day 4
Day 3
Day 0
Day 6-12
1. An egg cell is released from the ovary.
Embryo

3.10 *The journey of the fertilised cell.*

An embryo becomes a foetus

When the embryo has implanted in the uterus lining, cell division continues. During this time, cells also begin to become specialised to form different parts of the body. Between 8 and 10 weeks after fertilisation, its organs have formed and its heart has started to beat, the arms and legs begin to form and the embryo begins to resemble a baby (figure 3.11). After this time, the developing baby is called a foetus.

The foetus continues to develop in the uterus. It is surrounded by a watery fluid called **amniotic fluid**, inside the amniotic sac (figure 3.12). The amniotic fluid supports and protects the foetus.

Key terms

amniotic fluid: liquid that surrounds the foetus in the uterus.

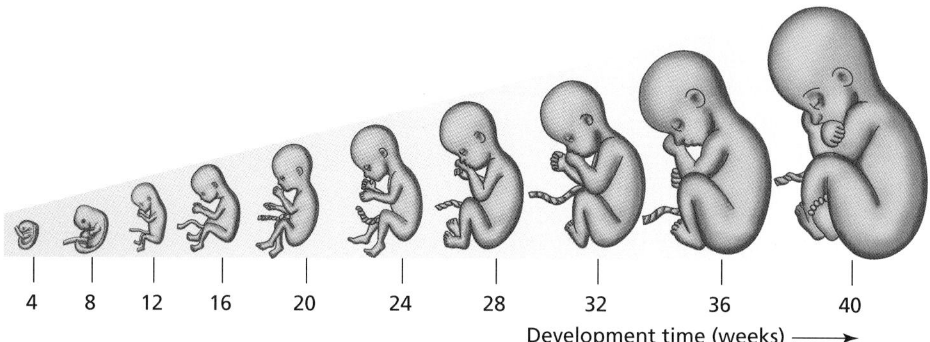

3.11 *Foetal growth from 4 to 40 weeks of development.*

Activity 3.3: Investigating the growth of the foetus

How much does a foetus grow at different times during development?

Pregnancy week	Mass (g)
8	1
12	14
16	100
20	300
24	600
28	1000
32	1700
36	2600
40	3500

This table shows the average mass of a foetus at different weeks of pregnancy.

A1 Draw a graph to show how the mass of the foetus changes during the different weeks.

A2 What is the increase in mass of the foetus between 12 and 16 weeks?

A3 During which weeks does the foetus 'grow' fastest?

A4 Between which weeks does the mass of the foetus double?

3 During which weeks do the legs and arms begin to grow, so the embryo resembles a baby?

4 What is the function of the amniotic fluid?

Activity 3.4: Investigating how to protect an egg

What is the best way to keep an egg safe from damage?

When a baby is growing inside the mother's uterus, the amniotic fluid plays an important role in protecting the baby. To see how important this fluid is, try the 'egg challenge'. You will need two raw eggs and two small plastic bags.

A Place one egg in a plastic bag, and seal the bag with a tight knot. Drop the egg onto a hard floor from a height of 20 cm. Observe what happens.

B Place the second egg in a plastic bag, fill the bag with water and seal tightly. Drop the water-filled bag onto the same hard floor from a height 20 cm. Observe what happens.

A1 What are the control variables (the things you kept the same) in this investigation?

A2 What happened in your investigation?

A3 What was the function of the water in this investigation?

A4 Use the results from this investigation to explain the function of the amniotic fluid in the uterus.

A5 Investigate further – at what height does the water-filled bag no longer provide enough support for the egg when you drop it?

The placenta

The **placenta** is an organ that grows from the embryo into the uterus lining during pregnancy. It connects the developing foetus to the uterus. The **umbilical cord** links the foetus to the placenta.

The baby's blood travels through the blood vessels in the umbilical cord and into the placenta. Figure 3.12 shows how close the baby's blood (in the placenta) comes to the mother's blood (in the uterus wall). This allows glucose, oxygen and water to diffuse from the mother to the foetus. Waste products, such as carbon dioxide, diffuse from the foetus to the mother.

Key terms

placenta: organ which forms in the uterus, linking the developing foetus to the uterus wall (and therefore the mother).

umbilical cord: flexible tube containing blood vessels from the foetus – it connects the foetus to the placenta (see figure 3.12).

Blood vessels from the umbilical cord and blood vessels in the uterus are very close to each other, but it is important to know that the mother's blood and the baby's blood *never mix.*

5 What is the placenta?

6 What substances exchange between the mother and the baby in the placenta?

7 What is the function of the umbilical cord?

8 Describe where diffusion happens in the placenta and why it is so important.

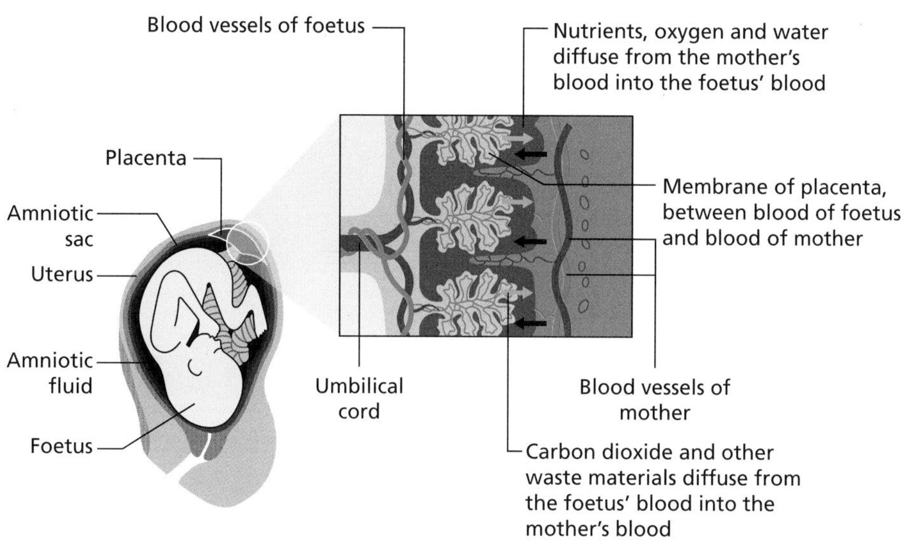

3.12 *A foetus developing inside the uterus. In the placenta, important materials exchange between the foetus' blood and the mother's blood, but their blood never mixes.*

Birth

The length of time that it takes for a foetus to fully develop in the uterus is the **gestation period**. The average gestation period for a human baby is 40 weeks (nine months), although this varies between different people. When birth begins, the mother starts to have contractions – the muscles in the wall of the uterus tighten and relax, again and again. The contractions help to get the baby in the correct position and to push it out of the mother's body through the vagina. The amniotic sac breaks during birth and the amniotic fluid flows out.

Before the baby can move out of the mother's body, the cervix muscles must relax and widen to allow the baby's head to pass through the vagina. The contractions help the cervix to widen.

Key term

gestation period: the amount of time for a baby to fully develop in the mother's uterus. This time varies for different animals – for example, 31 days for rabbits and 22 months for elephants.

The baby is pushed out of the uterus first and then the placenta follows. The placenta is still attached to the baby by the umbilical cord. Once the baby is born, the umbilical cord is cut.

9. Name *three* things that happen to allow a baby to be born.

10. Why does the uterus wall contract and relax?

Gestation period

The gestation period varies for different animals. For example, for rabbits it is only 31 days, whereas elephants are pregnant for 22 months. This picture shows a female orangutan and her baby, in Sumatra, Indonesia. An orangutan's gestation period is about nine months, the same as for humans.

3.13

Placental rituals

What happens to the placenta after birth? This varies depending upon the culture. In parts of Ghana, Vietnam and Bali, for example, some people feel the placenta is almost like a twin of the newborn baby. The placenta is cleaned, wrapped or placed in a container, and carefully buried in a special place outside the home. In many African cultures, the burial place of the placenta is called 'zan boko'.

Key facts:

- ✔ A fertilised egg develops into an embryo, which implants into the uterus lining.
- ✔ At around 10 weeks, the embryo has become a foetus.
- ✔ The mother's blood comes close to the foetus' blood in the placenta but they never mix.
- ✔ Substances diffuse between the mother's blood and the foetus' blood in the placenta.
- ✔ The amniotic fluid protects the foetus.
- ✔ During birth, the amniotic sac breaks, the cervix gets wider and the muscles in the uterus contract and relax to help push the baby out.

Check your skills progress:

I can plot a line graph.

I can interpret a line graph.

Drugs, disease and diet

Learning outcomes

- To describe different types of drugs
- To identify ways in which drugs, diet and diseases can affect human growth and development

Starting point

You should know that...	You should be able to...
Smoking may damage the human body	Collect and interpret information from secondary sources
Some microorganisms are harmful to humans	

Drugs

A drug is a chemical that changes the way your body works. Drugs can be used for many different purposes. Some drugs help to fight disease, some reduce pain and others are decongestants (drugs that can provide relief from a blocked nose). Drugs that are used for medical purposes are **pharmaceutical drugs**. Other drugs are used for enjoyment and many of these recreational drugs are **illegal** in many countries.

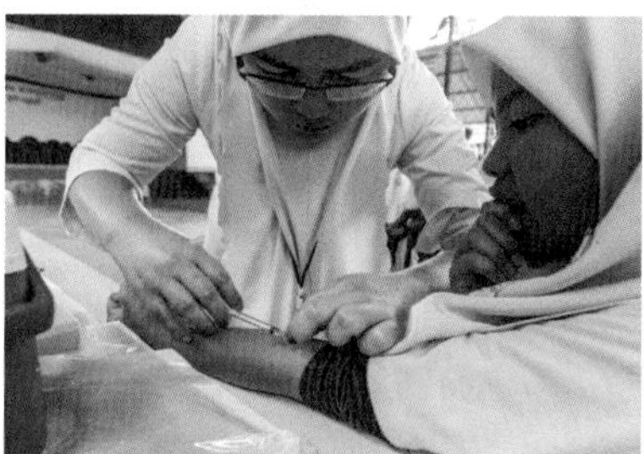

3.14

Because drugs affect how your body works, they may also change how it grows and develops.

1 Give *three* different uses for pharmaceutical drugs.

Key terms

illegal drug: drug that individual people are not allowed to buy or use. Different countries have different laws about drugs.

pharmaceutical drug: drug used in healthcare to help the body fight a disease, or to relieve pain.

Effects of alcohol

Alcohol is a type of drug. It is a depressant, which means that it slows down your body's responses. For example, it can slow down how long it takes someone driving a car to press the brake pedal in an emergency.

If a mother drinks large amounts of alcohol while she is pregnant, this can have serious effects on the baby's development. Alcohol passes from the mother's blood to the foetus' blood in the placenta (figure 3.12) and affects the growth of the foetus' cells.

If a mother drinks too much for a long time when she is pregnant, the baby may not grow very big. It may also have a smaller than average head and have problems with its sight and hearing.

2 What effect does alcohol have on a person's responses?

3 What damage can be caused to a developing foetus if the mother drinks alcohol excessively when pregnant?

4 Describe the path that alcohol takes through a woman's body to a foetus in her uterus.

The effects of tobacco smoke

When smoking tobacco, a person inhales over 4000 substances, most of which have harmful effects on the body. Smoking as an adult causes problems for many different organ systems, including the respiratory system.

If a woman smokes while she is pregnant, the substances in the smoke move into the blood of the developing foetus through the placenta. Two of these substances are carbon monoxide and nicotine (the drug in tobacco). They reduce the amount of oxygen that the foetus gets, which can have damaging effects.

Babies born to mothers who smoke during pregnancy may have low birth mass, reduced growth, underdeveloped lungs, heart defects and delayed brain development. Smoking during pregnancy also increases the chances of a **stillborn** baby (a baby that is dead when it is born).

5 Name *two* substances that are inhaled when a person smokes.

6 Give *three* ways in which a woman who smokes may damage her developing foetus.

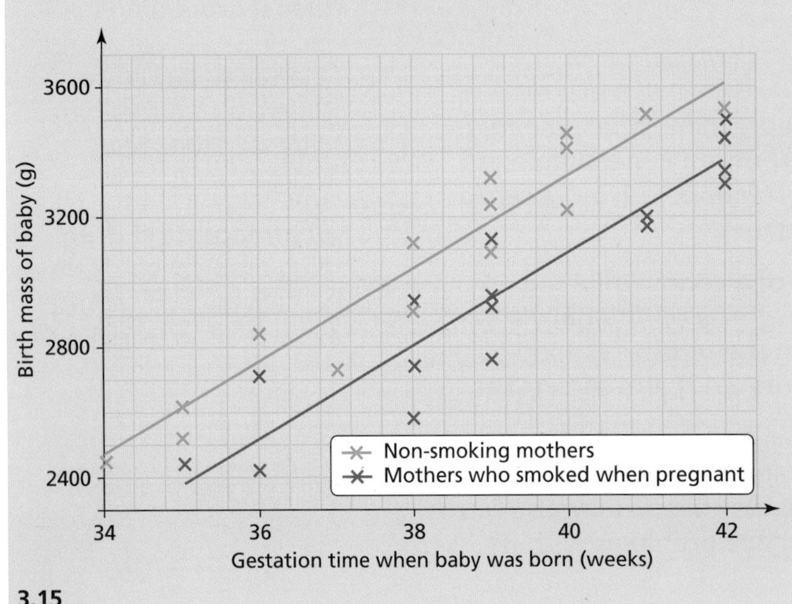

3.15

Key term

stillborn: the term used to describe a baby which is dead when it is born.

7. Look at the scatter graph in figure 3.15.
   a) Make a conclusion from the graph.
   b) Explain why the line for 'non-smokers' is above that of 'smokers'.

Drugs and pregnancy

If a mother takes drugs while she is pregnant, then the foetus is probably 'taking the drugs' too. This is because drugs often pass from the mother's blood to the foetus' blood in the placenta. Some drugs have harmful effects on the foetus, increasing the chance of birth defects, premature birth (before gestation is complete) and being stillborn.

Doctors usually avoid giving pharmaceutical drugs to pregnant women. An example of a drug that causes problems for developing babies is thalidomide.

In the late 1950s, thalidomide was prescribed for many pregnant mothers to prevent 'morning sickness' (sickness in pregnancy). When women who had taken this drug gave birth to babies with severe limb abnormalities (as well as other problems), doctors started to see that this drug could have terrible effects on the developing foetus.

8. If a pregnant woman takes drugs, what problems might this cause to a developing foetus?
9. Why did doctors stop prescribing thalidomide to pregnant women?

3.16 *This person was born with limb abnormalities caused by the drug thalidomide, which his mother was prescribed while she was pregnant.*

Activity 3.5: Researching the effects of drugs on a foetus' development

How can you inform people about the harmful effects of drugs in pregnancy?

Imagine that you are doctor. One of your jobs is to advise women who are trying to get pregnant about the risks of taking drugs.

A1 Research the impact of four different drugs on the development of a foetus.

A2 Create a leaflet or poster to give information to women about the dangers of taking drugs while they are pregnant.

The effects of disease

A person's growth and development can be affected by many different factors, including diet and disease. Diseases that affect the brain, such as a tumour, can change the amount of **growth hormone** that is made in the brain. This affects a person's growth.

Key term

growth hormone: chemical made in the brain that causes growth in the body.

The effects of diet

In order to grow, a person needs to have a diet that includes all the different nutrients. A person who does not have all the essential nutrients in their diet is unlikely to grow as fast as other people. Protein, for example, is essential for muscle growth and repair.

The impact of excess growth hormone

Sultan Kösen was born in 1982 in Turkey. He's 2 m 51 cm tall! Kösen grew to this height because of a tumour in his brain, which meant that he was making a lot more growth hormone than normal. So he kept on growing!

3.17 *Sultan Kösen.*

10 What factors might affect a person's growth and development?

11 Where in the body is growth hormone made?

Biofortification

3.18 *Biofortified sweet potatoes.*

Scientists around the world are developing 'biofortified' crops, which contain more vitamins or minerals than standard crops. People eat these biofortified foods to increase the amounts of certain vitamins and minerals that they have in their bodies, and to stop certain problems occurring.

One type of biofortified sweet potato contains increased levels of vitamin A. Our bodies need vitamin A for good vision. A deficiency of vitamin A can lead to eye damage and blindness, along with growth problems in children. But some people are concerned that biofortified crops may have unknown effects on health and the environment.

Key facts:

- ✔ Using certain drugs during pregnancy can cause serious damage to the foetus, including reduced growth, delayed brain development and heart defects. It also increases the risk of stillbirth.
- ✔ Some diseases can affect the growth and development of individuals.

Check your skills progress:

I can use information from secondary sources to explain scientific ideas to others.

End of chapter review

Quick questions

1. Which letter is labelling the nucleus of the egg cell?

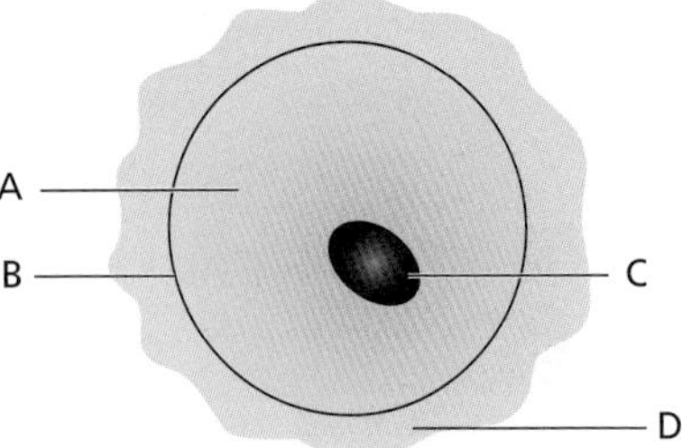

3.19 [1]

2. Why does an egg cell contain a food store in its cytoplasm?

   (a) To protect the egg

   (b) It contains nutrients for the developing embryo

   (c) It contains the information for making a new animal

   (d) It is a water reserve for the developing embryo [1]

3. This is a diagram of the female reproductive system.

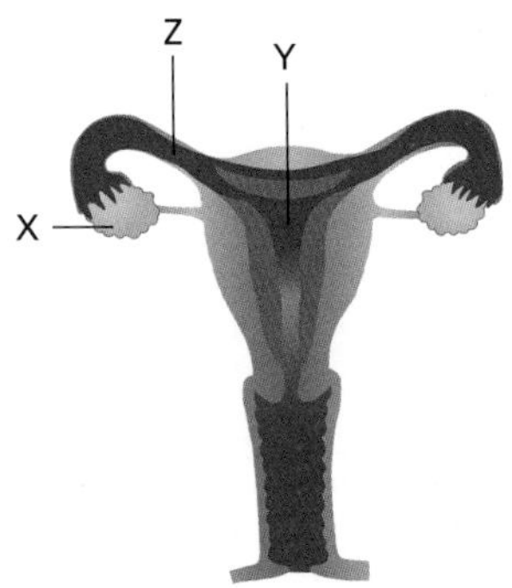

3.20

   (a) What is the name of the structure labelled X? [1]

   (b) What is the name of the structure labelled Y? [1]

   (c) What is the name of the structure labelled Z? [1]

4

Chapter 4

Elements, compounds and mixtures

The salt we use on our food has the chemical name sodium chloride. Seawater is a mixture of salt dissolved in water. Salt is separated from seawater using a change in state called evaporation.

In hot, dry countries like Thailand and Indonesia, salt is produced from shallow pools of seawater. The hot sun heats up the water so it evaporates. The salt is left behind.

You will learn about:

- How particle theory explains changes of state
- How particle theory explains gas pressure and diffusion
- What an element is and how elements are listed on the Periodic Table
- How compounds are different from elements and mixtures

You will build your skills in:

- Planning a method that can be used to test an idea
- Identifying hazards and planning to control risks

More on changes of state

Learning outcomes

- To use particle theory to explain the properties of solids, liquids and gases and what happens during changes in state

Starting point

You should know that...	You should be able to...
Substances exist as solids, liquids and gases. These are the three states of matter	Group materials as solids, liquids or gases
The particles in solids, liquids and gases move differently and are arranged differently	
When a substance is heated, the kinetic energy of the particles increases and this may cause a change in state	

Particle theory

You have already learned that all materials are made up of particles. This is a model called particle theory.

In some materials the particles are single atoms. In others, they are groups of atoms joined together to make molecules.

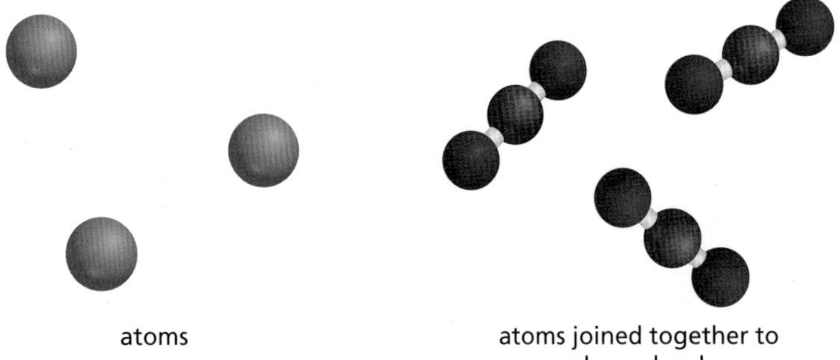

4.1 *Particles can be single atoms or molecules.*

When drawing diagrams using particle theory, each particle (whether an **atom** or a **molecule**) is always shown as a single sphere. This makes the model easier to use.

Key terms

atom: the smallest particle of a substance (element) that can exist and still be the same substance.

molecule: group of two or more atoms joined together. Oxygen, carbon dioxide and water all exist as molecules.

Explaining the properties of solids, liquids and gases

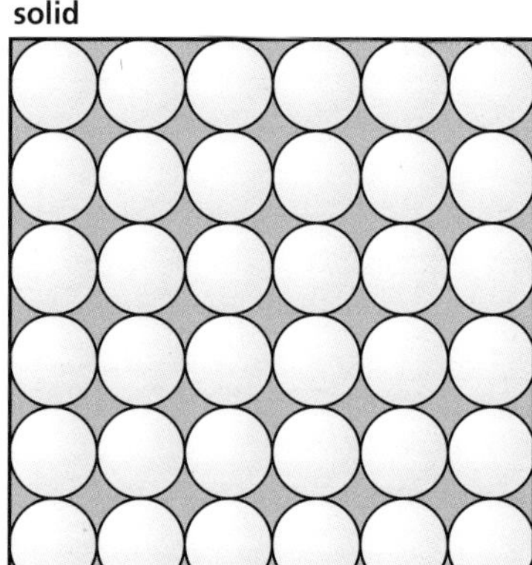

4.2 *The particles in a solid.*

You should remember that the particles in a solid are in fixed positions. They cannot move around but are vibrating in one place. The particles are held together by strong forces.

1 Use particle theory to explain why solids cannot flow or be compressed.

Not all solids have the same properties. You can tear a polythene bag, but not a sheet of iron. Some solids are stronger than others. We can explain this using particle theory. The forces that hold the particles together in iron are much stronger than the forces that hold the particles together in a polythene bag.

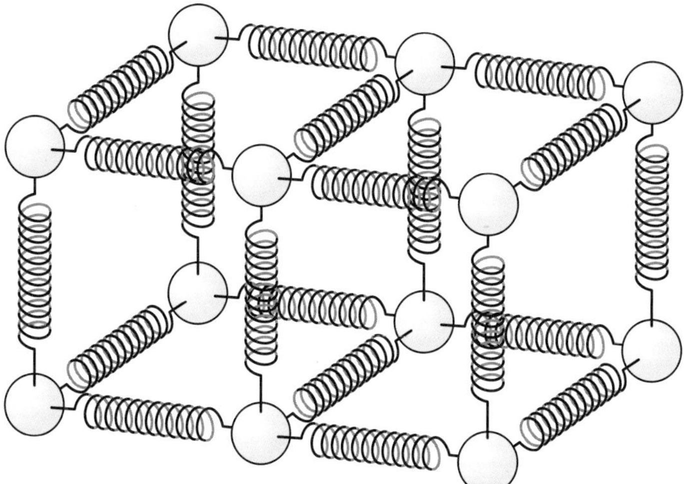

4.3 *We can model the forces between particles in a solid as springs.*

2 What are the strengths and weaknesses of this model?

3 Describe how you would change the model shown in figure 4.3 to show a solid with strong forces or one with weak forces.

The forces between the particles in a liquid are weaker than those in a solid. The particles are close together and touching. They are able to move around.

The forces between the particles in a gas are extremely weak. The particles are far apart and move around quickly and randomly. They hit each other and the sides of the container they are in.

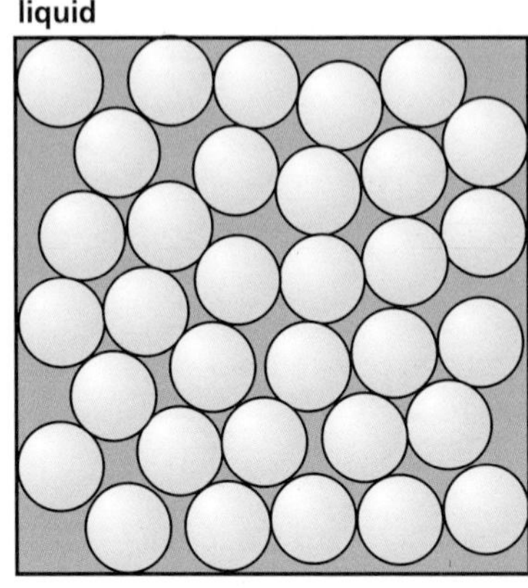

4.4 *The particles in a liquid.*

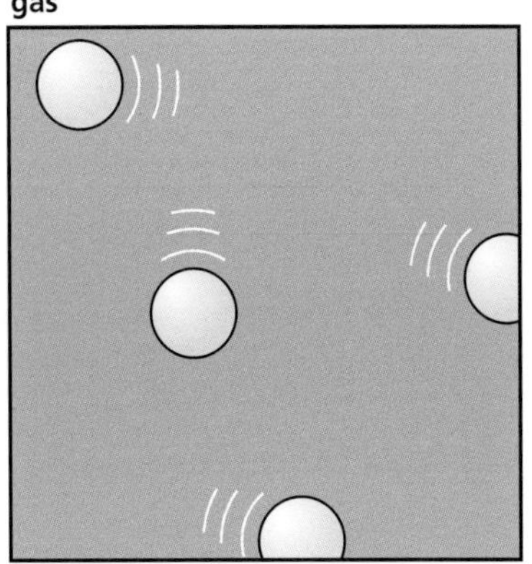

4.5 *The particles in a gas.*

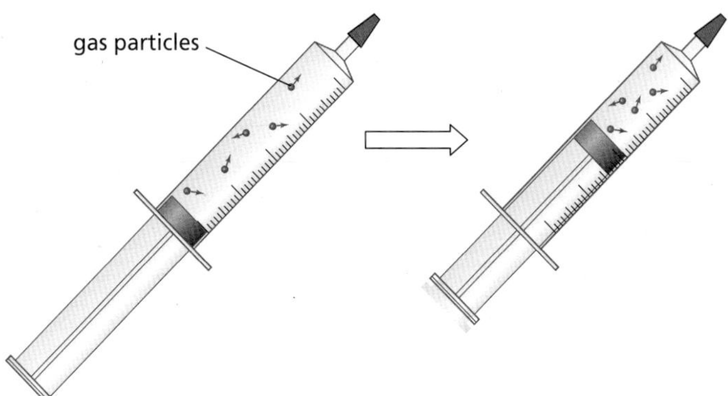

4.6 *Compressing a gas pushes the particles closer together.*

4 Use particle theory to explain why:

a) Liquids can flow.

b) Gases do not have a fixed shape or volume.

Explaining changes in state

You should remember that particle theory helps to explain what happens when a material changes state.

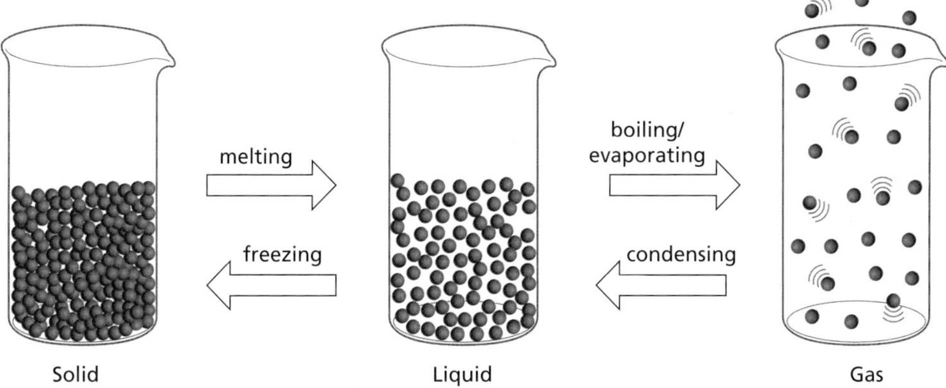

4.7 *When a material changes state the arrangement of its particles changes.*

When you heat a solid, heat (thermal) energy is transferred to its particles.

The particles start to vibrate more – their kinetic (movement) energy will increase. Some of the particles will vibrate so much that they break away from their fixed positions. Eventually, all the particles are moving around, but still touching each other. The material has melted and is now a liquid.

If you continue to heat the liquid, the kinetic energy of the particles will increase even more. Eventually they will have enough energy to move apart from each other. The material is now a gas.

Activity 4.1: Changes in state storyboard

A storyboard is a series of pictures that tells a story.

Create a storyboard to show what happens to the particles in a solid as it is heated to form a gas.

Use these questions to help you plan:

- How many pictures will you use?
- What will each picture show?
- How can you show how the movement of the particles changes?

Include text underneath the pictures to explain what they are showing.

5 Name the changes in state where:

a) The particles gain kinetic energy.

b) The particles lose kinetic energy.

6 Use particle theory to explain what happens when:

a) Steam cools down to form water.

b) Liquid water changes into ice

Solar in the Sahara

The Noor 1 power plant at the edge of the Sahara Desert in Morocco will help the country to supply most of its energy from renewable sources by 2030. The power plant contains half a million curved mirrors that focus the Sun's energy onto a pipe full of liquid, heating it to 393 °C. The hot liquid transfers heat energy to water, producing steam that turns electricity-generating turbines. The heat energy turns water from a liquid to a gas.

4.8 *The Noor 1 power plant covers an area the same as 200 soccer pitches. It is so big it can be seen from space.*

Key facts:

- ✔ There are forces between particles in solids, liquids and gases.
- ✔ The strength of the forces determines the properties of materials.
- ✔ When a material is heated, the kinetic energy of the particles increases and the motion and closeness of the particles changes. This might result in a change in state.

Check your skills progress:

I can communicate explanations clearly.

Gas pressure and diffusion

Learning outcomes

- To use particle theory to explain gas pressure and diffusion
- To explain how temperature affects the rate of diffusion

Starting point

You should know that...	You should be able to...
The particles in a gas and a liquid are moving around	Identify variables and choose the correct equipment to collect evidence to answer a question
When particles gain energy they move faster and further apart	Write a prediction using scientific knowledge
	Use evidence collected from an investigation to write a conclusion and say whether a prediction is correct, or not

Gas pressure

The particles in a gas are far apart and moving around. They collide with each other and the sides of the container they are in.

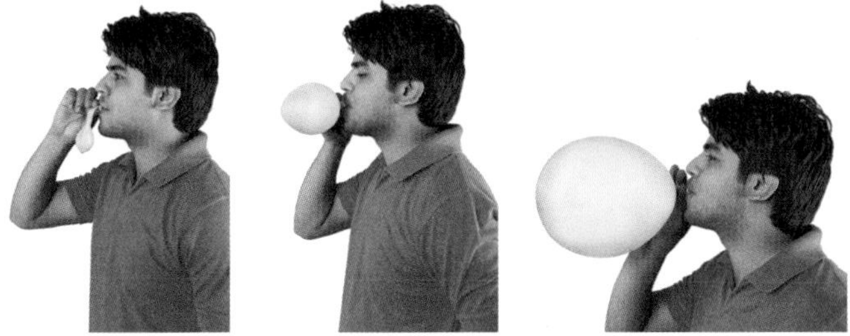

4.9 *When you blow into a balloon you are putting more air into it. The idea of gas pressure helps explain why the balloon gets bigger.*

It is difficult to squash (compress) a balloon.

This is because the gas particles inside are hitting the sides of the balloon. The force of the gas particles on the inside of the balloon is called **gas pressure.**

Key term

gas pressure: the effect of the forces caused by collisions from gas particles on the walls of a container.

Activity 4.2: Using particle theory to explain gas pressure

Your teacher will give you a container with some small balls in it. This is a model, and the balls represent gas particles.

A1 Shake the container so the balls move around.

They will hit each other and the sides of the container. These are collisions.

The force of the balls colliding with the sides of the container is what causes gas pressure.

A2 Use the model to show what happens to gas pressure when:

a) The number of gas particles in the container changes

b) The temperature of the gas changes

c) The size (volume) of the container changes.

Gas pressure is increased when there are more particles inside the container. This increases the frequency of collisions with the sides (how often collisions happen).

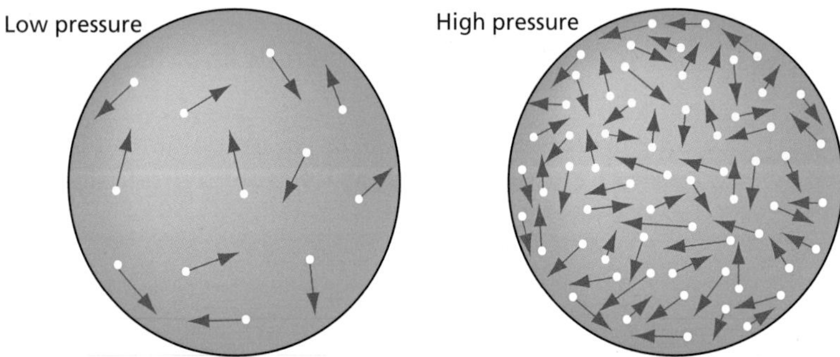

4.10 *Increasing the number of particles in the same space increases the gas pressure.*

Increasing the temperature also increases gas pressure. This is because the particles have more energy, so move about more quickly. They collide with the walls of their container more often and with more force.

1 Yasmin inflates a balloon.

a) Predict how the gas pressure inside the balloon changes as it is inflated.

b) Use particle theory to explain why.

2 Look back at figure 4.6. Predict what will happen to the gas pressure inside the syringe as the plunger is pushed down. Use particle theory to explain why.

3 Leo notices that his soccer ball feels much harder in the day, when it is hot, compared to at night when it is cooler. Use particle theory to explain why.

Diffusion

Particles spread out from a region where there are lots of them (a high **concentration**) to a region where there are fewer (a lower concentration). This is called **diffusion.**

Key terms

concentration: measure of how many particles of a certain type there are in a volume of liquid or gas.

diffusion: the spreading out of particles from where there are many (high concentration) to where there are fewer (lower concentration).

Diffusion happens in gases. If someone opens a bottle of perfume in a room, the perfume particles will diffuse through the air. Soon everyone in the room will be able to smell the perfume.

Diffusion can also happen in liquids. If you put a drop of red dye into a glass of water, after a few hours the dye particles will have spread out and mixed with the particles of water.

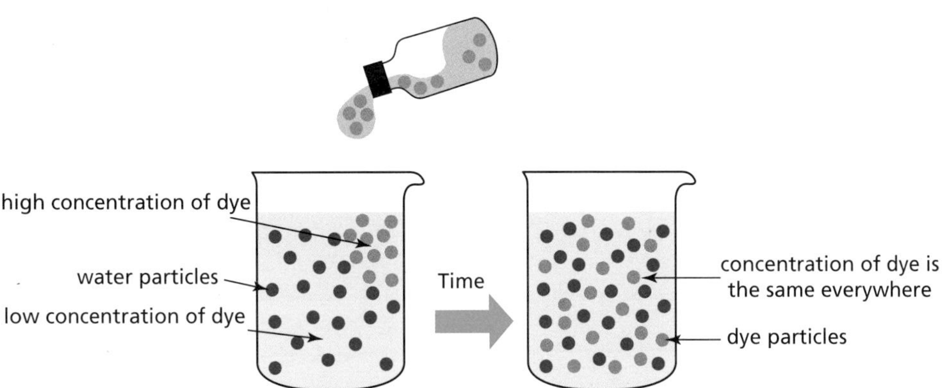

4.11 *In the drop of dye, the dye particles are at a high concentration. The dye particles diffuse in the water until their concentration is the same throughout.*

Activity 4.3: Why does diffusion happen?

Your teacher will give you a tray with some small balls in it. The balls are different colours.

A Separate them, so the balls of one colour are at one end of the tray and the balls of the other colour are at the opposite end.

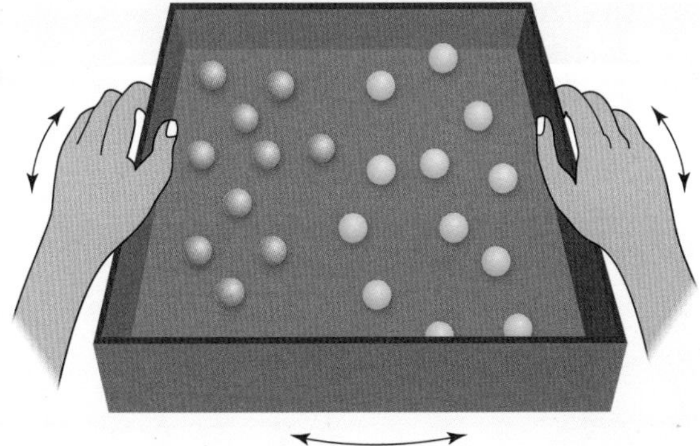

4.12

B Keep the tray on the table and shake it gently from side to side and up and down so the balls move around.

Questions to think about:

A1 What does shaking the tray model?

A2 What happens to the different coloured balls?

A3 How does this model show diffusion?

A4 Use the model to explain how diffusion happens.

Diffusion in a liquid or gas happens because particles are moving about randomly in all directions and collide with other particles.

When a collision happens, the particles move off in different directions. The particles get mixed up. Eventually they will be evenly spread out.

The problem with pollution

Air pollution is a big problem in many of the world's cities. Factories, power plants and vehicles release polluting gases and particles of smoke into the air. Wearing a face mask blocks the large particles of smoke but does not stop pollutant gases such as carbon monoxide and nitrogen dioxide from entering your lungs. This is because these particles are much smaller. These gases diffuse from your lungs into your blood, where they can travel around your body and harm organs.

4.13 *When pollutant gases diffuse from the lungs into the blood, they can harm health.*

What affects the rate of diffusion?

The **rate** of diffusion is how quickly it happens.

Look at the experiment in figure 4.14.

Both liquids evaporate and diffuse as gases through the tube. When the two gases meet they react to form a white solid.

You can see that the white solid does not form in the middle of the tube. This is because the mass of ammonia particles is less than the hydrochloric acid particles. They diffuse faster. The rate of diffusion is higher.

Key terms

rate: measurement of how quickly something happens.

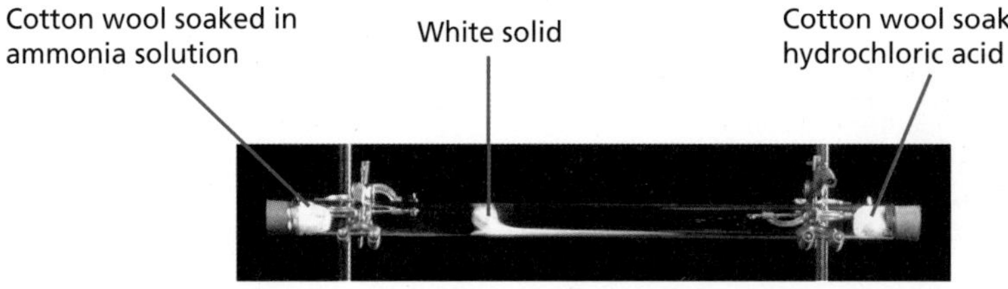

4.14 *This experiment shows that mass of particles affects the rate of diffusion.*

Activity 4.4: Does temperature affect the rate of diffusion?

When a coloured sweet is added to water, the dye dissolves in the water and diffuses away from the sweet.

Use this to plan an investigation to test the idea that temperature affects the rate of diffusion.

A1 Write your plan:

a) What variable will you change? How will you change this and what range will you choose?

b) What variable will you measure?

c) What variables need to be kept the same?

A2 Make a prediction: what do you think your results will show?

A3 Follow your plan and record your results in a table.

A4 Write a conclusion. Say what the results show and compare them with your prediction.

A5 Use particle theory to explain your results.

4 Use what you know about diffusion to explain why:

a) You can smell what is cooking in the kitchen from another room.

b) The colour from a teabag spreads throughout a cup of water.

c) The smell of a scented candle is much stronger when it is lit.

d) Diffusion happens more quickly in gases than in liquids.

5 The scents musk and ginger oil are both liquids. They are mixed together to make a perfume.

The forces between the particles of musk oil are stronger than the forces between the particles in ginger oil.

When the perfume is put on the skin both scents can be smelled. But, after a few hours only the musk can be smelled. Explain why.

Key facts:

- ✔ Gas particles are always moving and so collide with each other and the sides of their container.
- ✔ The force of gas particles colliding with their container is called gas pressure.
- ✔ The particles in liquids and gases spread out from where they are at a high concentration to where they are at a low concentration. This is called diffusion.
- ✔ The rate of diffusion is affected by temperature and the mass of the particles.

Check your skills progress:

I can plan an investigation to test an idea.

I can make predictions using scientific knowledge and understanding.

Chapter 4 . Topic 3

Atoms, elements and the Periodic Table

Learning outcomes

- To understand that elements are made of atoms
- To remember the symbols of the first 20 elements on the Periodic Table

Starting point

You should know that...	You should be able to...
All substances are made up of particles	Research data using secondary sources
Particles can be atoms or molecules	

Elements

Everything in the universe is made up of atoms. Atoms are very small. They are so small that it is difficult to imagine their size. Clench your hand up into a fist. If each atom in your fist was the same size as a marble, then your fist would be about the same size as Earth!

There are only around 100 different types of atom.

Materials that just contain only one type of atom are called **elements**. Examples of elements are silver, carbon and helium.

Key term

element: substance that contains only one type of atom; it cannot be split into anything simpler.

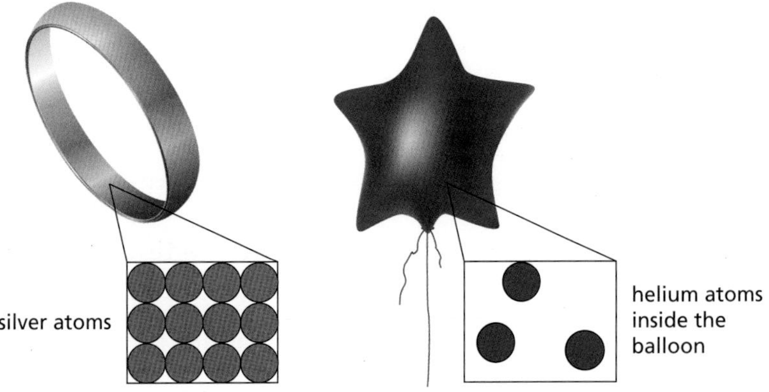

4.15 *Silver and helium are both elements. Silver contains only silver atoms. Helium contains only helium atoms.*

Using indium

Indium is a rare element but demand is high and supplies are decreasing. Indium is a metal. When indium is mixed with two other elements, tin and oxygen, it forms a material that has very special properties. It is transparent and can conduct electricity so is used to make touchscreens on smart phones, tablets and computer screens.

When old electronic devices are thrown away the indium is lost. However, Chinese scientists have found a way to recover indium from the compounds used on the screens. This recycling means that we should have enough indium to continue making new electronic devices for a long time.

4.16 *The element indium is used to make touchscreens.*

Activity 4.5: Exploring elements

Your teacher will give you some elements to observe.

A1 Describe its colour, whether it is a solid, liquid or gas and if it is a metal or non-metal.

A2 Put the elements into groups: how many different types of grouping can you use? Be prepared to explain why you chose these groups.

A3 Choose *one* element and use secondary sources to find out some more information, for example: What is it used for? What is its boiling and melting point?

Symbols and the Periodic Table

The name of an element is not the same in every language.

sølv	വെള്ളി	argent	**silber**	**ασήμι**
серебряный	রুপা	银	ચાંદીના	चांदी
銀	zilver	வெள்ளி	เงิน	چاندی

4.17 *The element name 'silver' in different languages.*

To make it easier for scientists around the world to talk about elements, each one has a **chemical symbol.**

The symbols for elements are the same in every language. The symbol for silver is Ag.

All elements that exist are shown with their symbols on the **Periodic Table**.

Key terms

chemical symbol: short way of representing an element's name.

Periodic Table: list of all the elements.

1 H																	2 He
3 Li	4 Be											5 B	6 C	7 N	8 O	9 F	10 Ne
11 Na	12 Mg											13 Al	14 Si	15 P	16 S	17 Cl	18 Ar
19 K	20 Ca	21 Sc	22 Ti	23 V	24 Cr	25 Mn	26 Fe	27 Co	28 Ni	29 Cu	30 Zn	31 Ga	32 Ge	33 As	34 Se	35 Br	36 Kr
37 Rb	38 Sr	39 Y	40 Zr	41 Nb	42 Mo	43 Tc	44 Ru	45 Rh	46 Pd	47 Ag	48 Cd	49 In	50 Sn	51 Sb	52 Te	53 I	54 Xe
55 Cs	56 Ba	57 La	72 Hf	73 Ta	74 W	75 Re	76 Os	77 Ir	78 Pt	79 Au	80 Hg	81 Tl	82 Pb	83 Bi	84 Po	85 At	86 Rn
87 Fr	88 Ra	89 Ac															

4.18 *The Periodic Table lists all the elements.*

The elements are arranged in rows from left to right. The first element is hydrogen, which has the symbol H. The next element is the next one along in the first row, which is helium (He).

Here are the names and symbols for the first 20 elements.

Name	Symbol
Hydrogen	H
Helium	He
Lithium	Li
Beryllium	Be
Boron	B
Carbon	C
Nitrogen	N
Oxygen	O
Fluorine	F
Neon	Ne

Name	Symbol
Sodium	Na
Magnesium	Mg
Aluminium	Al
Silicon	Si
Phosphorus	P
Sulfur	S
Chlorine	Cl
Argon	Ar
Potassium	K
Calcium	Ca

4.19 *Carbon is element number 6. It is found in lots of different forms, such as diamond.*

1. Use the Periodic Table to name the element number 14.
2. Which of these statements are true?
   a) Aluminium only contains one type of atom.
   b) The symbol for sodium is So.
   c) Oxygen contains two types of atom.
   d) The symbol for potassium is K.

4.20 *Sulfur is element number 16. It is a yellow solid often found in active volcanoes.*

Key facts:

- ✔ Elements only contain one type of atom.
- ✔ Each element has a symbol.
- ✔ Elements are arranged in the Periodic Table.

Check your skills progress:

I can remember the symbols of the first 20 elements in the Periodic Table.

Compounds and formulae

Learning outcomes

- To describe the difference between an element and compound
- To identify hazards and plan how to control risks when carrying out an experiment

Starting point

You should know that...	You should be able to...
Elements only contain one type of atom	Choose the right equipment for an experiment and use it correctly
Each element has a symbol; these symbols are shown on the Periodic Table	Recognise hazard warning labels
A reversible change is a change in a substance that can be changed back again	

Compounds

There are only around 100 different elements but there are many millions of different materials.

This is possible because atoms of different elements can join together to form substances called compounds.

The atoms in a **compound** are strongly held together. It is difficult to separate them.

Some examples of compounds are water, carbon dioxide and sodium chloride (salt).

You already know that each element has its own symbol. Every compound has a chemical **formula**.

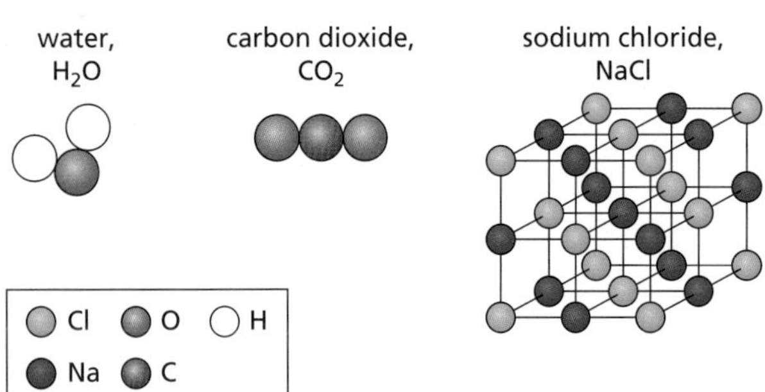

4.22 *The names and chemical formulae of some compounds, and how the atoms are joined together.*

The symbols in the formula show what elements are in the compound.

The number after a symbol in the formula shows how many of that type of atom there are.

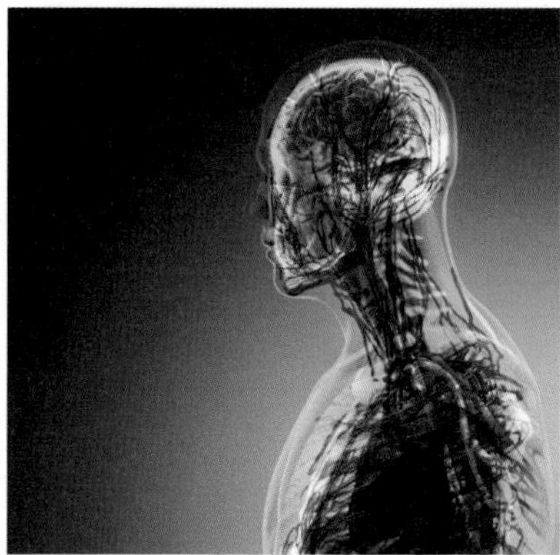

4.21 *Almost all of your body is made up of just six elements: oxygen, carbon, hydrogen, nitrogen, calcium and phosphorus. These elements are able to form many materials with different properties like bone, blood, skin, hair and nails because they combine in different ways to form compounds.*

Key term

compound: contains atoms of more than one element strongly held together. Compounds have different properties to the elements they contain.

The formula of water is H_2O. Each molecule of water contains two hydrogen atoms and one oxygen atom.

The formula of sodium chloride is NaCl. Sodium chloride does not contain molecules. It exists as a crystal. A crystal of sodium chloride (NaCl) contains an equal number of sodium and chlorine atoms.

Key term

formula: shows the chemical symbols of elements in a compound, and how many of each type of atom there are.

1. Describe the difference between an element and a compound.
2. How many carbon and oxygen atoms are in one molecule of carbon dioxide (CO_2)?
3. A molecule of glucose has the formula $C_6H_{12}O_6$.
   a) What elements make up glucose?
   b) How many of each atom are in one molecule of glucose?
4. The compound lithium oxide has the formula Li_2O.
   Which statement is correct?
   a) There is the same number of lithium atoms as oxygen atoms.
   b) There is double the number of lithium atoms as oxygen atoms.
   c) There is half the number of lithium atoms as oxygen atoms.
5. The formula of magnesium chloride is $MgCl_2$.
   A piece of magnesium chloride contains 400 chlorine atoms. How many magnesium atoms are there?

Making compounds

Magnesium is an element. It is a light grey metal.

Oxygen is an element. It is a colourless gas found in the air.

When magnesium is heated in the air it glows bright white.

Magnesium atoms join with oxygen atoms to form the compound magnesium oxide. This is a change called a **chemical reaction.** The magnesium oxide cannot easily be separated into the elements it contains because the magnesium and oxygen atoms are strongly held together.

The properties of magnesium oxide are different to the properties of the elements it is made from.

Key term

chemical reaction: a change in which new substances are produced.

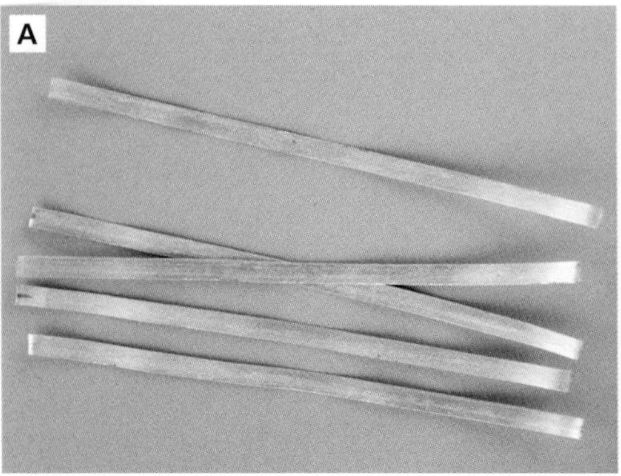

4.23 *When a strip of magnesium (A) is heated in air (B) it glows bright white and forms a white solid (C).*

6 Describe how the properties of magnesium oxide are different to the elements it is made from.

7 A teacher heats together iron and sulfur. Iron sulfide is produced.

a) Name *two* elements mentioned.

b) Name the compound.

c) This is a chemical reaction. Explain what this means.

Identifying risks

When you carry out an experiment there might be hazards. These are things that can cause harm. You might see **hazard** labels on the substances you use.

A **risk** is the chance of a hazard causing harm to you, or people around you.

One step in an experiment might be to pour some hot water into a test tube. Before you start you should identify the hazards and plan how to control the risks (reduce the chances of harm). This is shown in table 4.2.

Key terms

hazard: harm that something may cause.

risk: chance of a hazard causing harm.

	Hazards	**Controlling the risks**
Hot water	Can cause burns if it touches your skin	Put the test tube into a test tube rack. Do not hold it.
Glass test tube	The test tube could break and cause cuts if touched	Tell the teacher if anything breaks. Do not pick up any broken glass yourself.

Table 4.2 Identifying hazards and controlling risks for an experiment.

Activity 4.6: Identifying risks

Samir is going to carry out a chemical reaction and make a compound called magnesium oxide.

He plans to hold a strip of magnesium in a Bunsen burner flame until it gives out a bright white light and magnesium oxide is made. He will then put the magnesium oxide onto his desk so he can observe it.

This experiment could cause harm to Samir and the people around him.

A1 Describe the hazards, and what Samir should do to reduce the chance of harm.

Key facts:

- ✔ Compounds are formed when atoms of two or more elements combine in a chemical change.
- ✔ Each compound has a formula, which shows you how many of each atom are in the compound.
- ✔ The properties of a compound are different to the elements it contains.

Check your skills progress:

I can identify hazards.

I can plan to control risks in investigations.

Separating mixtures

Learning outcomes

- To explore how to separate mixtures based on their composition
- To plan a way of separating a mixture

Starting point

You should know that...	You should be able to...
When solids do not dissolve or react with water, they can be separated by filtering	Choose what equipment is relevant to use in a particular situation
Some solids dissolve in water to form solutions	Identify hazards
When a liquid evaporates from a solution the solid is left behind	Plan how to control risks in investigations

Making mixtures

A **pure** substance contains only one type of element or compound. For example, iron is a pure substance (an element). Pure water contains only water molecules (a compound).

A **mixture** is made up of at least two different elements or compounds. For example, water and salt together make a mixture.

Here are some other examples of mixtures.

4.24 *The air is a mixture of different gases, mainly nitrogen and oxygen.*

4.25 *This milkshake is a mixture of milk, sugar and strawberries.*

4.26 *This rock is a mixture of different compounds called minerals.*

The substances in a mixture can easily be separated because the different substances are not strongly held to each other.

1 Describe the difference between a compound and a mixture.

2 Air contains the gases nitrogen, oxygen, water vapour and carbon dioxide.

Key terms

mixture: two or more elements or compounds mixed together. They can easily be separated.

pure: substance that contains only one element or compound.

4.27 *A mixture of small pieces of iron and sulfur can easily be separated using a magnet because iron is magnetic and sulfur is not.*

For each of these substances, write down whether it is an element, a compound or a mixture:

a) Air

b) Oxygen

c) Nitrogen

d) Carbon dioxide

e) Water

3 Mo is making concrete. He mixes sand, small stones and cement powder. He then adds water. Before he adds the water, it is a mixture. After he adds the water it is not.

Explain why.

Activity 4.7: Element, compound or mixture?

Your teacher will give you some different substances.

Each one will have some information about it. For example, its name or formula.

A1 Use your observations and the information to write down if each one is an element, compound or mixture.

A2 Explain to a partner how you made your decision.

Separating mixtures

Mixing together substances is a physical change; it can easily be reversed.

There are many methods of separating mixtures. Each uses different equipment. The method you choose will depend on the substances in the mixture.

When you separate a mixture, it is important to use the equipment carefully and correctly so the substances you produce are pure.

Here are some examples of how to separate mixtures.

Using filtration

Adding sand to water produces a mixture. The sand is **insoluble**.

To separate a mixture of an insoluble solid and a liquid you use **filtration.**

Filter paper contains tiny holes. The water particles are small enough to pass through the holes.

The sand particles are too large so they stay in the funnel.

Key terms

filtration: method used to separate an insoluble solid from a liquid.

insoluble: substance that does not dissolve.

Using evaporation

Adding salt to water produces a mixture.

The salt is **soluble**. It dissolves in the water to produce a **solution**. You cannot see the salt, but it is still there.

To separate a soluble solid from a solution you use **evaporation.**

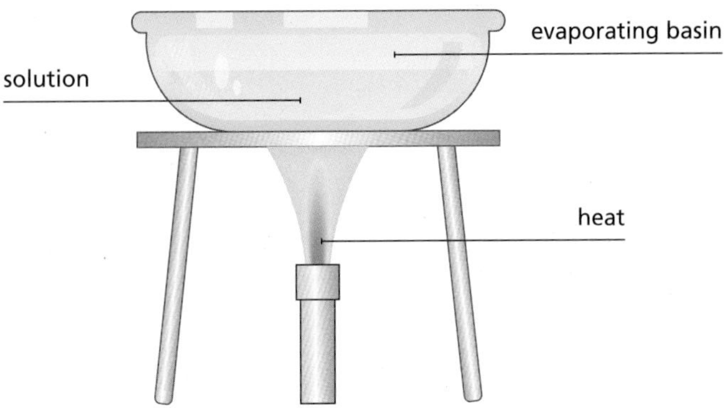

4.29 *The equipment used for separating a mixture using evaporation.*

The solution is heated so the water evaporates.

The salt cannot evaporate. It is left in the basin.

Using distillation

When you heat a solution the evaporated liquid mixes with the air.

If you want to collect the liquid from a solution you use **distillation**.

In figure 4.30, salty seawater is distilled to form pure water.

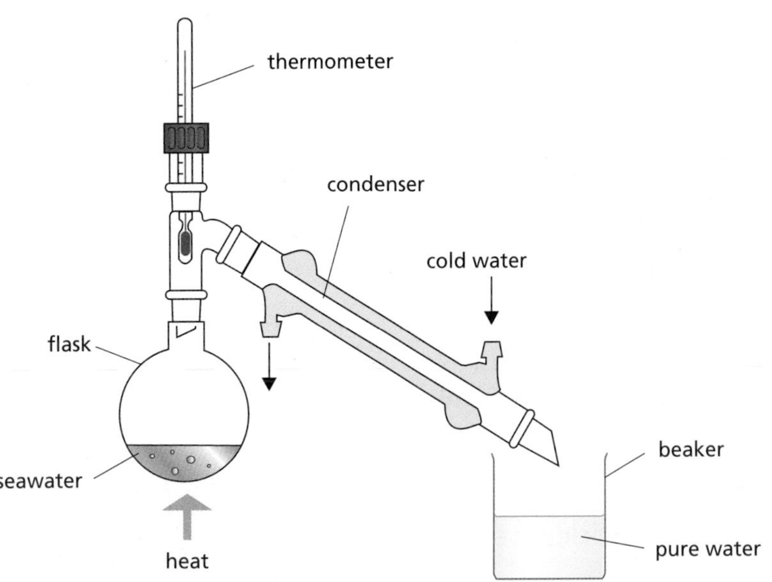

4.30 *The equipment used for separating a mixture using distillation.*

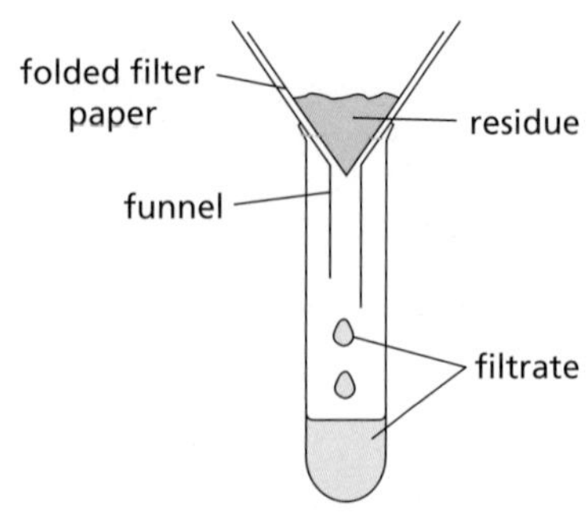

4.28 *The equipment used for separating a mixture using filtration.*

Key terms

distillation: separation method used to separate a liquid from a mixture.

evaporation: method used to separate a soluble solid from a liquid.

soluble: substance that dissolves to form a solution.

solution: mixture of a soluble substance and a liquid.

The salt solution is heated so the water boils. The thermometer shows that this happens at 100 °C.

The steam passes through the condenser. The walls of the condenser are cold so the steam condenses.

The liquid water passes out of the condenser and is collected in a beaker.

The salt is left in the flask.

4 For each of these mixtures, choose a method to separate them.

a) Sugar and water

b) Sand and water

5 You can use both evaporation and distillation to separate salt and water.

Choose which method you would use to make salt from seawater.

Give a reason for your choice.

6 Explain the function of the cold water in the distillation equipment.

7 The boiling temperature of ethanol is 78 °C. The boiling point of water is 100 °C.

Describe how the distillation equipment in figure 4.30 can be used to separate a mixture of ethanol and water.

Explain how the method works.

Supplying fresh water

1.2 billion people, nearly a sixth of the world's population, live in areas where it is difficult to supply people with enough water for drinking, cooking and washing.

Separating pure water from seawater by distillation is one way of solving this problem. In huge distillation plants seawater is heated until it boils. The steam is then cooled and condensed to form pure water, which is stored in large tanks before being taken to people's homes.

4.31 *Salty seawater is distilled to form pure water in Dubai.*

Activity 4.8: Separating a mixture

Your teacher will give you a mixture.

A1 Plan a way of separating the mixture. You need to choose the correct equipment.

A2 List the hazards to yourself and to others in the class and suggest how to reduce the risks.

A3 Follow your plan and separate the mixture. You should use the equipment correctly to make sure the substances you produce are pure.

Key facts:

- ✔ A mixture is made up of at least two different elements or compounds.
- ✔ The substances in a mixture can easily be separated.
- ✔ Separation methods include filtration, evaporation and distillation.

Check your skills progress:

I can plan a method and suggest appropriate safety procedures.

I can choose equipment and use it correctly.

End of chapter review

Quick questions

1. In which change of state do the particles lose energy?

   a boiling b evaporating c melting d condensing **[1]**

2. Describe how the movement and arrangement of particles change when a liquid freezes. **[2]**

3. The spreading out of particles from a region where there of lots of them to a region where there are fewer is called...

   a gas pressure b diffusion c melting d condensation **[1]**

4. Give the symbols for the following elements:

   (a) Helium **[1]**

   (b) Sodium **[1]**

   (c) Chlorine **[1]**

5. Which *two* factors affect the rate of diffusion?

   a mass of the particles b temperature of the particles

   c colour of the particles d shape of the particles **[2]**

6. How many hydrogen atoms are in a molecule of each of these compounds?

   (a) Water (H_2O) **[1]**

   (b) Methane (CH_4) **[1]**

   (c) Hydrochloric acid (HCl) **[1]**

7. Which *one* of these is an element?

   a carbon b carbon dioxide c air d water **[1]**

8. Copper is a shiny metal. A piece is heated in oxygen. Its surface becomes dull and black.

   (a) Name the compound made in this chemical change. **[1]**

   (b) Describe how the compound is different to copper. **[2]**

9. A chemical reaction between hydrochloric acid and magnesium produces magnesium chloride ($MgCl_2$).

   Is magnesium chloride an element, compound or mixture? **[1]**

Connect your understanding

10. Some air is pumped into a bicycle tyre.

(a) Predict what will happen to the amount of gas pressure inside the tyre as air is pumped in. **[1]**

(b) Use particle theory to explain your prediction. **[2]**

11. Use particle theory to explain why you can compress a gas but not a liquid. **[3]**

12. Rock salt is found in the ground. It is a mixture of rock particles and salt.

Describe how you would produce pure, dry salt from rock salt.

You should explain the function of each step. **[4]**

13. Rashin added a few drops of blue ink to a glass of cold water.

(a) Describe what she will see in the glass. Explain why. **[2]**

(b) She then added a few drops of the ink to a glass of hot water. Describe how her observations would be different to what happened in the glass of cold water. Give a reason for this. **[3]**

14. Draw diagrams to show the particles in:

(a) a piece of solid copper. **[2]**

(b) carbon dioxide gas (CO_2). **[2]**

Challenge question

15. In humans and other mammals, air is breathed into the lungs.
Oxygen travels by diffusion from air in the lungs into the blood.

The blood must keep moving round the body to keep the rate of diffusion high.
Explain why. **[3]**

5

Chapter 5

Metals, non-metals and corrosion

What's it all about?

Leaving your bicycle outside is not a good idea. Many parts are made from steel, which rusts when it reacts with oxygen and water in the air.

This is an example of a chemical reaction that is not useful. The rust stops parts of the bicycle from moving.

You will learn about:

- The differences between metals and non-metals
- How the physical properties of metals make them a useful material
- Why some chemical reactions are not useful

You will build your skills in:

- Planning an investigation to test an idea
- Identifying important variables; choosing which to change, control and measure
- Making predictions using scientific knowledge and understanding

Metals and non-metals

Learning outcomes

- To describe and explain the differences between metals and non-metals

Starting point

You should know that...	You should be able to...
Materials can be grouped as metals or non-metals based on their physical properties	Outline plans to carry out investigations, considering the variables to control, change or observe
Metals are usually hard, strong, malleable and good conductors of heat and electricity	
Some materials are magnetic, but many are not	

Metals and non-metals on the Periodic Table

You have already learned that all the elements are listed in the Periodic Table. Elements are either metals or non-metals.

Elements with similar properties are grouped together.

The metal elements are grouped on the left side of the Periodic Table. The non-metal elements are grouped on the right.

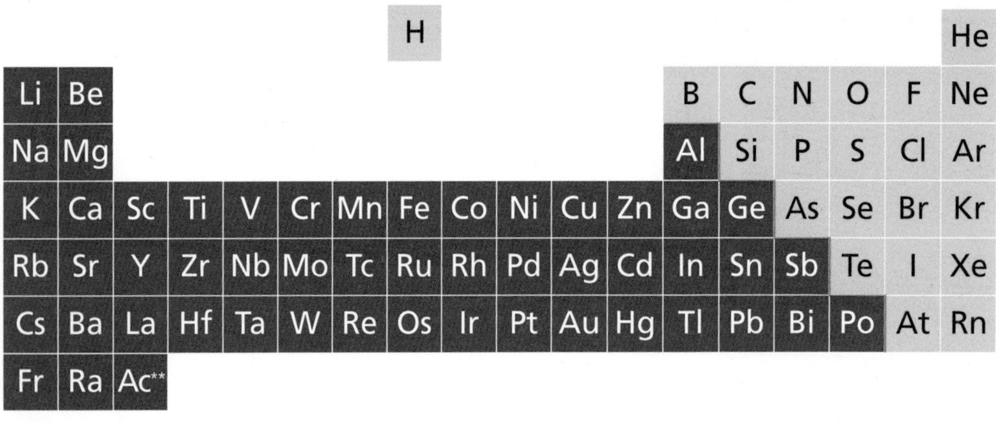

5.1 *Metals and non-metals are in different places on the Periodic Table.*

1 For each of these element symbols, name the element and say if it is a metal or non-metal:

a) Be

b) Cl

c) Al

d) Ar

2 List the symbols and names of *three* other metals shown in figure 5.1.

Mixtures and compounds

Alloys are mixtures of metals. They are made by heating two or more metals together.

No chemical reaction takes place, so the different metals in an alloy are not chemically joined together.

Alloys often have different properties from the elements they are made from, giving them different uses. Bronze is an alloy of the elements copper and tin. Bronze is much harder than the elements it is made from.

If a metal reacts with a non-metal a compound is formed. The compound is a non-metal.

The metal and non-metal atoms in the compound are chemically joined, so it is difficult to separate them.

Key term

alloy: mixture of metal with other elements.

5.2 *Bronze has been used for thousands of years to create statues.*

Physical properties of metals and non-metals

Most metals have similar physical properties.

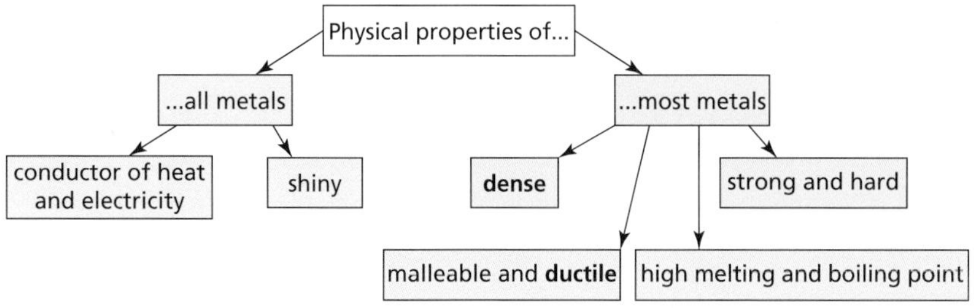

Most non-metals have different properties to metals.

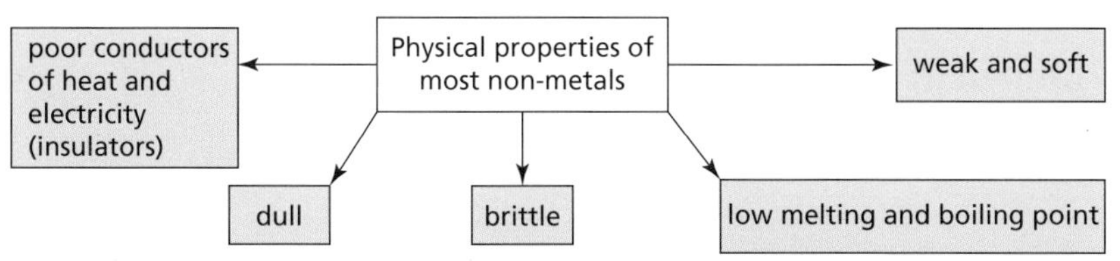

5.3 *Sapphires are made of the compound aluminium oxide. This is formed when aluminium reacts with oxygen.*

5.4 *Sulfur is a non-metal that exists as a solid at room temperature. It is dull (not shiny), brittle (breaks easily when bent) and soft (easily made into a powder).*

3 Use the information to decide if the material is a metal or non-metal.

A: A shiny material that conducts electricity.

B: A dull, brittle material.

C: A gas at room temperature.

D: A malleable material that is solid at room temperature.

Explaining physical properties

Iron is a metal and sulfur is a non-metal. The physical properties of both can be explained by thinking about how their atoms are arranged, and the forces between the atoms.

Melting and boiling points

Iron has a very high melting point. This means that it is a solid at room temperature. You need to heat iron to 1538 °C to melt it.

This is because the atoms in metals are held together by very strong forces. A lot of energy is needed to separate the atoms so they are free to move, and form a liquid. Iron also has a very high boiling point.

Sulfur is a non-metal. Its atoms are arranged in molecules and the forces holding the molecules together are much weaker than the forces between the atoms in iron. It has a much lower melting point than iron.

5.5 *The melting point of sulfur is 115 °C. It can be melted easily using the heat from a Bunsen burner.*

Strength and hardness

Iron is both strong and hard.

This is because of the strong forces between the atoms. A lot of energy is needed to break apart the structure of a metal.

Sulfur is soft and weak, it can easily be ground into a powder.

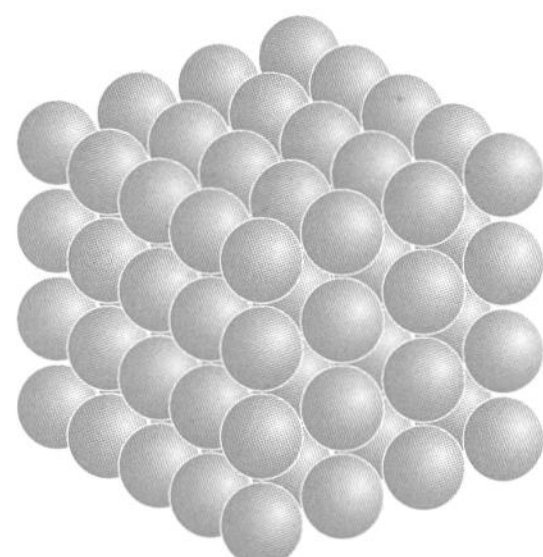

5.6 *The atoms in iron are arranged in a closely packed, regular pattern.*

Density

Iron is **dense**. It has a high density. Its atoms are arranged in a regular pattern and are close together. This means many iron atoms fit in a small volume.

The molecules in sulfur are not arranged in a regular pattern so they are not as close together as the atoms in iron. Sulfur has a lower density than iron.

Changing shape

Sulfur is brittle – it will break when bent; but iron is very malleable – it can be bent and shaped. This is because the atoms in metals are arranged in layers. The layers can slide over each other if a large force is applied.

Because layers of particles are able to slide past each other, metals can also be pulled to form long, thin wires – they are **ductile**.

Key terms

dense: has a high mass in a small volume.

ductile: able to be stretched into wires.

4 The table shows the melting points of elements in row 3 of the Periodic Table.

Symbol	Na	Mg	Al	Si	P	S	Cl	Ar
Melting point (°C)	98	639	660	1410	44	113	−101	−189

a) Name the *two* elements that are gases at room temperature (25 °C).

b) Explain why they have low melting points.

c) Most non-metals have low melting points. Name the non-metal in the table that has a high melting point.

d) Sodium (Na) has quite a low melting point. Describe why this is unusual.

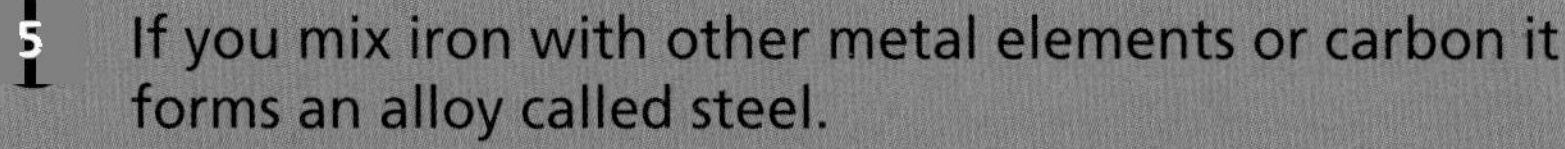

5 If you mix iron with other metal elements or carbon it forms an alloy called steel.

This changes the regular structure of the pure metal. In the diagram, the different colours show the different atoms in steel.

5.8

Use the diagram to suggest why steel is less malleable than pure iron.

6 Research the melting points of the first 20 elements. Plot a bar chart with a suitable scale to show the melting points. (Remember that some elements have melting points below 0 °C.)

5.7 *When a force is applied to a metal the layers of atoms slide over each other.*

Uses of metals and non-metals

Metals and non-metals both have useful physical properties. They have many different uses. The type of metal or non-metal is carefully chosen based on its properties.

Copper is a very good conductor of electricity. It is also ductile.

Helium is a gas at room temperature. It has a very low density even for a gas, so will float in air.

5.9 *Copper is a good choice for making electrical wires.*

7 An engineer is choosing a material to make the blades on a jet engine.

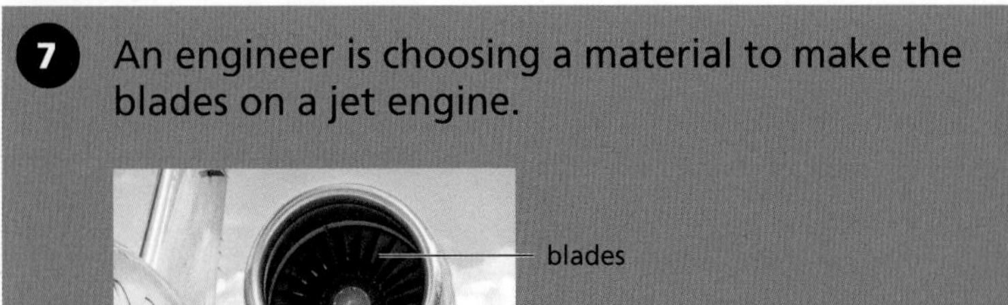

5.11

Suggest what physical properties the material must have. Give reasons for your choices.

5.10 *Helium-filled balloons float in air.*

Activity 5.1: Investigating the strength of metal wires

Some scientists wanted to find out which metal was the strongest.

They planned to test a short length of metal wire using heavy masses.

They will add masses to the hanger until the wire snaps.

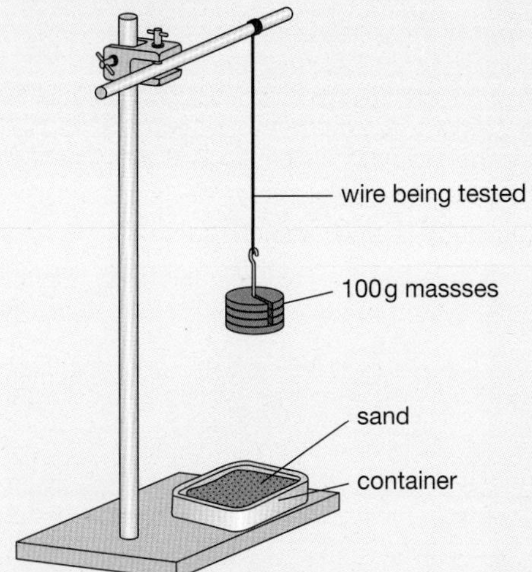

5.12

Help them to plan their investigation.

A1 What variable should they change?

A2 What variable should they measure?

A3 What variables must they control (keep the same)?

A4 How should they control risks to themselves and others when carrying out this investigation?

A home from home

The International Space Station (ISS) is a home for astronauts from all over the world who live and work together. The materials it is built from are very different to homes back on Earth.

5.13 *The materials used to make the ISS were carefully chosen to reduce its mass but keep it strong.*

The ISS had to be lifted into orbit from Earth so aluminium was chosen to build the structure because it has a lower density than other strong metals, like steel. Aluminium has a low density because atoms of aluminium have a lower mass than atoms of iron and are also less tightly packed.

The ISS must also provide protection to the astronauts from impacts by tiny meteoroids and synthetic debris. Aluminium is quite a soft metal, so layers of Kevlar, ceramic fabrics and other non-metal materials form a blanket up to 10 cm thick around the aluminium shell to strengthen it and protect the astronauts inside.

Unusual metals

Some metals have unusual properties.

Most metals are grey or silver in colour. Copper and gold are not.

Most metals are hard and have high melting points. Sodium is soft; it can be cut with a knife. It also has a low melting point.

Most metals have a high density. Potassium has a low density. It can float on water.

Iron, cobalt and nickel are the only metal elements that are **magnetic**.

All metals are solid at room temperature except mercury, which is a liquid.

5.14 *Gold is chosen to make jewellery because of its colour.*

5.15 *Mercury is the only metal that is a liquid at room temperature.*

Key term

magnetic: material that is attracted by a magnet.

8 Non-metals can also have unusual properties.

For each of these non-metals, say which property is unusual (there might be more than one).

a) Graphite is a form of carbon. It is dull, soft, can conduct electricity and has a higher melting point than any metal element (3600 °C).

b) Silicon is a shiny, brittle insulator.

c) Diamond is another form of carbon. It is a good conductor of heat and is very hard.

Key facts:

- ✔ Metals are found on the left side of the Periodic Table and non-metals are on the right.
- ✔ Most metals are shiny, hard, strong and dense. They have high melting and boiling points and can conduct heat and electricity.
- ✔ Most non-metals are dull, soft and brittle. They have low melting and boiling points and are insulators.
- ✔ Many of the properties of metals and non-metals can be explained by the way their particles are arranged and the strength of the forces between them.

Check your skills progress:

I can identify important variables; choose which variables to change, control and measure.

Chapter 5 . Topic 2

Corrosion and rusting

Learning outcomes

- To describe corrosion and rusting as chemical reactions which are not useful
- To plan an investigation to test an idea

Starting point

You should know that...	You should be able to...
A chemical reaction is a change where new substances are produced	Outline plans to carry out investigations
Compounds are formed when atoms of two or more elements combine in a chemical reaction	Write a prediction using scientific knowledge
The properties of a compound are different to the elements it contains	Present results in the form of tables and use them to make conclusions

Chemical reactions

Many chemical reactions are useful.

Chemists use chemical reactions to manufacture a huge variety of new substances, like medicines and plastics.

In vehicle engines, fuel reacts with oxygen to release energy which is used for movement.

In plants, a chemical reaction between carbon dioxide and water makes food molecules such as sugars and starch.

5.16 *Bread rises because yeast use a chemical reaction to produce a gas (carbon dioxide) that forms bubbles inside the dough.*

How do metals change?

Metals can change over time. They get damaged because of chemical reactions. This is called **corrosion**.

Corrosion happens because metals react with substances in the air or water to form new compounds.

This can make the metal weaker and brittle.

Key term

corrosion: the damage of metals through chemical reactions with substances in the air or water.

5.17 *Iron drain pipes can become corroded which weakens them. In time, they may break and leak.*

Rusting

Iron, and its alloy steel, are very useful metals. For example, steel is strong so it is used to build bridges.

But, there is one problem with iron and steel – they rust.

Rusting is the corrosion of iron and iron alloys. It happens because iron reacts with oxygen in the air, when water or water vapour is present.

Rusting is an example of an **oxidation** reaction. Rust is a red/brown substance that contains compounds of iron and oxygen. It is much softer and weaker than iron.

1. Give some uses of iron and steel.
2. Explain why rusting is not a useful chemical reaction.

5.18 *The hole in this steel car is caused when the steel turns into rust, which breaks up.*

Key terms

oxidation: chemical reaction with oxygen to form a compound that contains oxygen.

rusting: chemical reaction of iron with oxygen and water. It is the corrosion of iron.

Activity 5.2: What conditions are needed for rust to form?

Tarek and Nour set up an experiment to find out what is needed for an iron nail to rust.

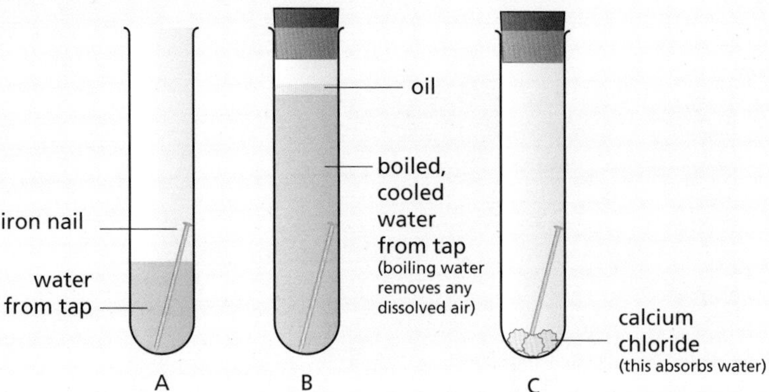

5.19

A1 Describe how they changed the conditions in each tube.

A2 Make a prediction: what do you think will happen to the iron nail in each tube: will it rust or not rust? Give a reason for your answer.

A3 Describe a method to investigate how the mass of an iron nail changes as it rusts.

A4 Give a prediction for this investigation, with a scientific reason.

Speeding up and slowing down

Rusting is a chemical reaction. The speed, or rate, of a chemical reaction can be changed.

The rate that iron rusts can be increased or decreased.

Activity 5.3: Investigating the rate of rusting

Samir notices that a scratch in the paint on his boat meant the metal underneath went rusty very quickly.

He thinks it could be because of the warm temperature or the salt in the seawater.

A1 Write a plan that you can use to investigate how *one* of these variables (temperature or the presence of salt in water) affects how quickly an iron nail rusts.

Things to think about:

a) What variable will you change?

b) How will you change it? What range will you use?

c) What will you measure to see how the variable has affected how much the iron has rusted?

d) What variables should you control?

A2 Design a suitable results table.

How can you prevent rusting?

To stop an iron object rusting you must prevent water and oxygen reacting with the iron.

There are different ways of doing this.

<table>
<tr><td>Painting

5.20 Steel car bodies are painted.</td><td>Greasing

5.21 This steel cable on a ship is greased to stop it from rusting.</td></tr>
<tr><td>Galvanising

5.22 Steel used for buildings is often galvanised. This means covering the steel with a thin layer of zinc.</td><td>Plastic coating

5.23 Chairs for use outdoors are sometimes coated in plastic.</td></tr>
</table>

3 Explain how these methods prevent rusting.

4 Choose a method to prevent each of these objects from rusting. Give a reason for your choice.

a) Underground metal pipes

b) A decorative garden gate

5 Painting and greasing could both be used to protect a nail from rusting. Write a method that you could use to investigate which method is best.

Strengthening concrete

Concrete is made stronger (reinforced) by adding steel bars to it. Reinforced concrete is used to build many structures, including buildings and bridges. Unfortunately, water and oxygen can move through the concrete. They react with the steel inside, which rusts. Rusting weakens the structure so it has to be replaced.

In the Middle East, the climate and environment mean that the steel rusts very quickly. The weather is hot and humid and the atmosphere has a high salt level. To slow down rusting, the steel bars are coated with a thin layer of very hard plastic. Even though this makes the concrete more expensive, it means it lasts longer.

5.24 *Plastic coated steel bars in a concrete column, for a bridge in Bahrain.*

Key facts:

- ✔ Metals can be damaged by chemical reactions with substances in air and water. This is called corrosion.
- ✔ Corrosion makes metals weaker. It is an example of a chemical reaction that is not useful.
- ✔ Rusting is the corrosion of iron. Oxygen and water must be present for iron to rust.
- ✔ Rust can be prevented by a barrier that prevents air and water from reaching the surface of the iron or steel.

Check your skills progress:

I can plan an investigation to test an idea.

I can identify important variables; choose which variables to change, control and measure.

I can make a prediction using my scientific knowledge and understanding.

I can write a conclusion to say what my results show.

End of chapter review

Quick questions

1. For each of these elements, use a Periodic Table to say whether it is a metal or non-metal.

   (a) magnesium **[1]**

   (b) neon **[1]**

   (c) nitrogen **[1]**

   (d) barium **[1]**

2. Silver is a typical metal.

   State *three* physical properties it has. **[3]**

3. Aluminium is used to build aeroplanes.

   Which *two* of its physical properties make it suitable for this use?

   a good conductor of electricity

   b malleable

   c ductile

   d low density **[2]**

4. Iron is magnetic.

   Name *one* other magnetic element. **[1]**

5. What *two* substances are needed for iron to rust?

   a oxygen

   b nitrogen

   c water

   d salt **[2]**

6. What type of chemical reaction is rusting?

   a neutralisation

   b oxidation

   c burning

   d heating **[1]**

7. Which *one* of these is a non-metal?

   a steel

   b bronze

   c sodium

   d diamond **[1]**

8. The hull (bottom) of a ship is built from steel.

   (a) Give *one* reason why steel is used to build ships. **[1]**

   (b) Give *one* problem with using steel to build ships. **[1]**

Connect your understanding

9. Element X is a gas at room temperature.

   Is element X a metal or non-metal? Give a reason for your answer. **[2]**

10. Bridges are usually made from steel, which rusts.

    (a) Explain why bridges are made from steel, even though it rusts. **[4]**

    (b) Describe *one* way of protecting a steel bridge from rusting. Explain how it works. **[3]**

11. Describe *one* advantage of using paint to prevent an iron gate rusting and *one* disadvantage. **[2]**

12. Iron nails can be galvanised. This means they are covered with a thin layer of zinc.

    Explain why this is used. **[3]**

13. Kashif and Olivia are investigating how the pH of water affects how quickly iron nails rust.

    They will place a small amount of water of different pH in the bottom of some test tubes.

    They will add an iron nail to each tube then leave them for 3 days and then observe how rusty the nails are.

    (a) Describe how they could change the pH of the water and measure it. **[2]**

    (b) List *three* variables they should control. **[3]**

Challenge question

14. Miriam carries out an experiment.

The equipment she uses is shown below. Iron wool is very thin strands of iron.

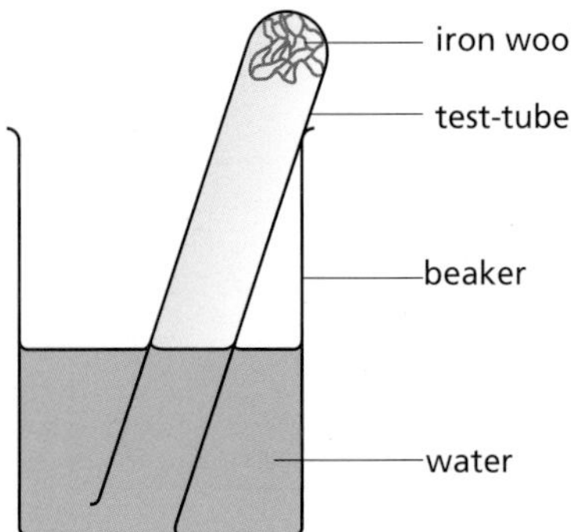

5.25

She leaves it for 5 days.

Predict what will happen to the iron wool and the level of water in the test tube.

Give scientific reasons for your predictions. **[5]**

6

Chapter 6

Chemical reactions

What's it all about?

From lift-off to colourful explosions of light, many different chemical reactions take place after a firework is lit.

To lift the firework a powder burns. This reaction releases gases very quickly in a very fast reaction. The colours come from compounds called salts whose atoms release coloured light when heated. Different metals in the salts give different colours: sodium salts produce yellow sparks and copper salts produce blue.

You will learn about:

- How to work out the name of a compound from the elements in it
- How to write word equations to describe chemical reactions
- How different salts can be made by reactions with acids
- The compounds formed when metals and non-metals react with oxygen

You will build your skills in:

- Identifying correlations in results
- Asking a scientific question that can be tested by carrying out investigations and collecting evidence

Using word equations

Learning outcomes

- To describe reactions using word equations

Starting point

You should know that...	You should be able to...
Some chemical reactions are useful and others (like rusting) are not useful	Identify hazards and plan to control risks in investigations
Compounds are formed when atoms of two or more elements combine in a chemical change	
Each compound has a formula, which shows you how many of each atom are in the compound	
An insoluble substance does not dissolve	

Chemical reactions

In a chemical reaction the substances that react are called **reactants**. New substances called **products** are made in the reaction.

There are clues that you can look for that show a chemical reaction might have happened. For example:

- Bubbles being given off. This shows that a gas has been made.
- A change in colour. A new product has been made which is a different colour to the reactants.

6.2 *When a copper coin is dipped into silver nitrate solution a chemical reaction takes place and the coin changes colour.*

- A change in temperature. This could be an increase or decrease in temperature.

Key terms

product: substance made during a chemical reaction.

reactant: substance that changes in a chemical reaction to form products.

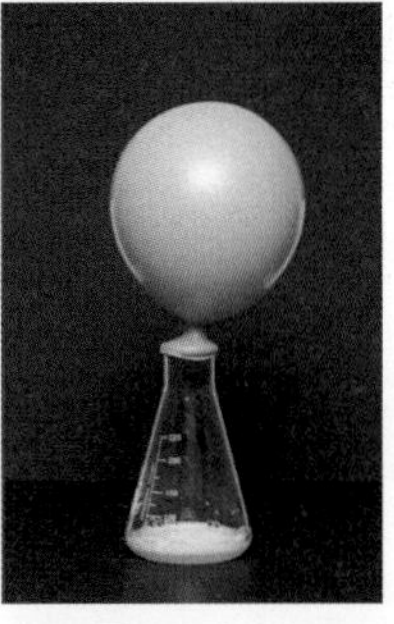

6.1 *Vinegar reacts with baking powder to produce bubbles of gas.*

6.3 *The reaction between iron and chlorine releases heat energy.*

One example of a chemical reaction is rusting.

The reactants are iron, water and oxygen. The product is rust. Iron is a grey metal, rust is brown. This change in colour is evidence that a chemical reaction has taken place.

Writing word equations

A **word equation** shows us the reactants and products in a reaction. The reactants and products can be elements or compounds.

You have already learned that magnesium and oxygen react together in an oxidation reaction. The product is magnesium oxide. This can be shown in a word equation:

magnesium + oxygen → magnesium oxide

Reactants Product

Word equations always have the reactants on the left. The arrow points to the products that are made.

1. A teacher reacts sodium and chlorine together. Sodium chloride is made. Write a word equation to show this reaction.
2. Photosynthesis is a chemical reaction that takes place in plant leaves. Carbon dioxide and water are used to make oxygen and glucose. Write this as a word equation.

Naming compounds

There are some rules to follow to help you to name a new compound formed in a chemical reaction.

- If the compound contains a metal, the name of the metal comes first.
- When two elements react to form a compound, the name often ends in *-ide*. For example, iron **chloride** and calcium **oxide**.

If a compound ends in *-ate* it also contains oxygen. For example, copper **carbonate** contains copper, carbon and oxygen atoms. Calcium **sulfate** contains calcium, sulfur and oxygen atoms.

Metal **hydroxide** compounds only contain atoms of the metal, hydrogen and oxygen.

Key term

word equation: model showing what happens in a chemical reaction, with reactants on the left of an arrow and products on the right.

Key terms

carbonate: compound that reacts with an acid to give carbon dioxide, a salt and water. For example, calcium carbonate ($CaCO_3$).

chloride: salt that is formed when hydrochloric acid reacts with another element; for example, sodium chloride (NaCl).

hydroxide: compound that contains one atom each of oxygen and hydrogen bonded together; for example, potassium hydroxide (KOH).

oxide: compound that is formed when oxygen reacts with another element; for example magnesium oxide (MgO).

sulfate: salt that is formed when sulfuric acid reacts with another element; for example copper sulfate ($CuSO_4$).

Table 6.1 shows some examples.

Name of compound	Metal element	Non-metal element/s	Formula
sodium chlor<u>ide</u>	sodium	chlorine	NaCl
magnesium ox<u>ide</u>	magnesium	oxygen	MgO
potassium <u>hydroxide</u>	potassium	oxygen, hydrogen	KOH
calcium carbon<u>ate</u>	calcium	carbon, oxygen	$CaCO_3$
lithium sulf<u>ate</u>	lithium	sulfur, oxygen	Li_2SO_4
silver nitr<u>ate</u>	silver	nitrogen, oxygen	$AgNO_3$

Table 6.1 *The name of a compound usually shows what atoms it contains.*

3 Write the name of each of these compounds. Use the Periodic Table to help you find the name of the element from its symbol.

a) LiCl

b) CaO

c) MgS

d) $NaNO_3$

e) K_2CO_3

Activity 6.1: Reacting sodium with water

A teacher showed his class the reaction of sodium with water.

He filled a large glass container with water and put in some universal indicator. He put some plastic screens between the container and the class.

He put on his splashproof eye protection and used a sharp knife to cut off a very small bit of sodium from a larger chunk.

He picked up the small bit using tweezers and added it to the water.

6.4

As soon as the sodium hit the water it gave off a gas which pushed it around the surface of the water. The reaction also gave out heat. The water turned purple.

A1 The products of the reaction are hydrogen and sodium hydroxide. Write a word equation for the reaction.

A2 Sodium hydroxide dissolves in water and is an alkali. What evidence is there that sodium hydroxide was made?

A3 List the hazards of doing this experiment.

A4 Explain how the teacher controlled the risks to himself and the class.

Rearranging atoms

During a chemical reaction the atoms in the reactants rearrange to form the products.

No atoms are lost. No new atoms are made.

In this chemical reaction, colourless silver nitrate solution is reacting with colourless sodium chloride solution. One product formed is white silver chloride, which is insoluble. It is a white solid and sinks to the bottom of the beaker. The word equation is:

silver nitrate + sodium chloride → silver chloride + sodium nitrate

Chemical reactions can also be shown as particle diagrams. They show you what happens to the atoms. The particle diagram for this reaction is:

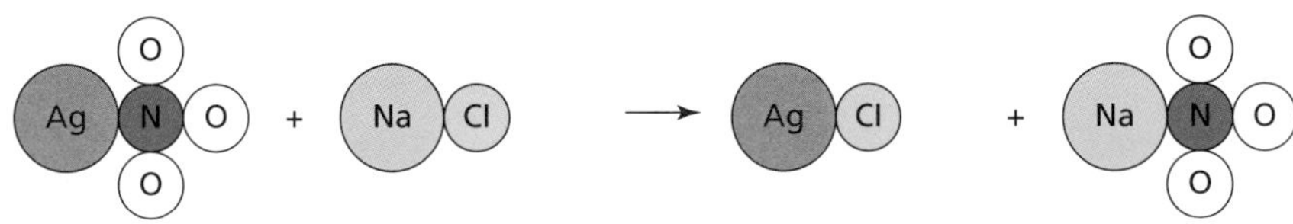

6.6

You can see that there are seven atoms on each side of the equation. No atoms have been lost or gained during the reaction.

The atoms of the elements in the reactants are also present in the products.

6.5 *Some chemical reactions result in a colour change.*

4 Name the *two* products made when silver nitrate reacts with sodium chloride.

5 Sunil predicts that because a solid is formed in the reaction, the mass of the products is more than the mass of the reactants. Do you think he is correct? Use the particle diagram to help you. Give a reason for your answer.

6 There are two products in the reaction between silver nitrate and sodium chloride but you can only see one. Explain why.

Removing mercury

Mercury is released into the environment through processes such as burning fossil fuels and mining. It can form soluble mercury compounds and dissolve in water. Mercury is very toxic and can cause brain damage so it is important that it is removed from drinking water. This can be done by adding aluminium sulfate to the water. Aluminium sulfate reacts with the soluble mercury compounds to form an insoluble mercury compound that can be filtered out of the water.

6.7 *Polluted water containing mercury can enter rivers from factories and mines.*

Key facts:

- ✔ During a chemical reaction, reactants react to form new products.
- ✔ A word equation separates the substances that react (on the left) and the products that are formed (on the right) with an arrow.
- ✔ Simple compounds are named using rules.
- ✔ No atoms are made or lost during a chemical reaction; they are just rearranged.

Check your skills progress:

I can identify hazards.

I can plan to control risks in investigations.

Reactions with acid

Learning outcomes

- To describe and write word equations for the formation of chloride and sulfate salts using reactions with acids
- To describe how to test a gas to see if it is hydrogen or carbon dioxide

Starting point

You should know that...	You should be able to...
Acids are substances that have a pH lower than 7, bases are substances that have a pH above 7	Use a word equation to describe a chemical reaction
Neutralisation is a chemical reaction which happens when an acid and an alkali or base react together	Name some simple compounds
There are signs that are evidence that a chemical reaction is taking place	Recognise results and observations that fit into a pattern

Reactions between metals and acid

If you add pieces of the metal magnesium to hydrochloric acid you will observe bubbles. The magnesium will get smaller and smaller and the outside of the test tube will feel warm.

The word equation for this reaction is:

magnesium + hydrochloric acid → magnesium chloride + hydrogen

The word equation shows you that hydrogen is a product. Hydrogen is a gas so it forms bubbles in the liquid. Nearly all metals react with acid to produce hydrogen gas. However, sometimes the reaction is very slow.

A **salt** is a compound formed when an acid reacts with a base or a metal. In this reaction the salt is called magnesium chloride. It is soluble so it dissolves to form a solution.

The reaction can also be shown as a particle diagram.

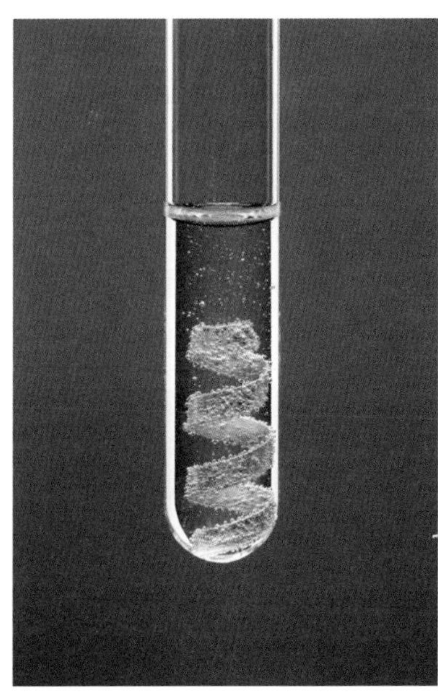

6.8 *Magnesium reacts with hydrochloric acid.*

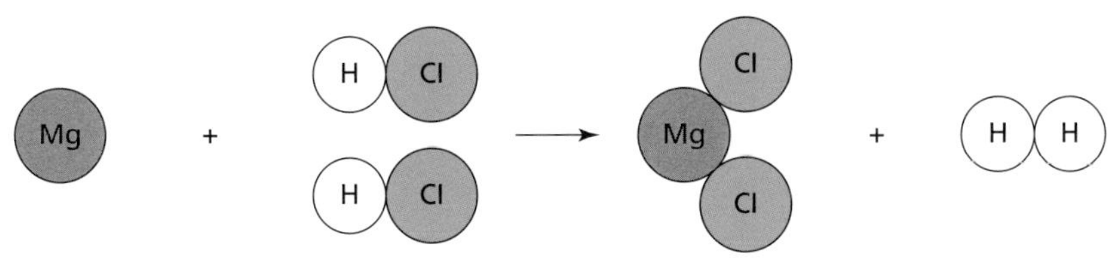

6.9 *The particle diagram shows how hydrogen gas is formed in the reaction between hydrochloric acid and magnesium metal.*

Key term

salt: compound formed when an acid reacts with a base or a metal.

During the reaction the hydrogen atoms joined to the chlorine atoms in the hydrochloric acid molecules separate. Two hydrogen atoms join together to form molecules of hydrogen gas. The magnesium atoms combine with the chlorine atoms to form a new compound; the salt magnesium chloride.

1 Describe the *two* signs that show magnesium is reacting with the hydrochloric acid.

2 Use the idea of particles to explain why the pieces of magnesium get smaller and smaller.

Testing for hydrogen

You can test the gas made in a reaction to see if it is hydrogen. Hydrogen is a flammable gas so if you put a lighted splint into the top of the test tube in which hydrogen is produced, you will hear a 'pop' sound. This is because the hydrogen burns very quickly.

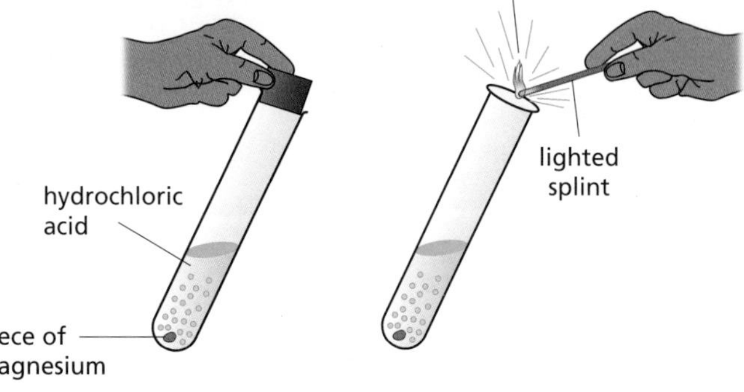

6.10 *How to test for hydrogen.*

Activity 6.2: Looking for patterns

A group of students investigated how different metals reacted with hydrochloric acid.

They put 5 cm^3 of acid into four test tubes and measured the temperature of the acid. They then added pieces of metal of the same size to the test tubes. They measured the highest temperature the acid reached.

They recorded their observations and results in a table.

Metal	Observations	Temperature of acid at start (°C)	Highest temperature reached (°C)
Magnesium	Fast bubbling. The magnesium got smaller quickly.	22.3	54.4
Copper	No bubbles.	24.4	24.5
Iron	A few bubbles. The iron did not change size.	23.0	26.8
Zinc	Some bubbling. The zinc got a bit smaller.	24.2	34.9

A1 Describe how the reactions are similar and how they are different.

A2 State what *two* variables were controlled.

A3 Metals can be ordered by their **reactivity**. More reactive metals will react more quickly with the acid. Order these metals, the most reactive first.

A4 Describe any **correlation** you can see between the reactivity and the temperature rise.

Making different salts

The salt made in a reaction depends on the reactants you use.

The first part of the salt's name is always the name of a metal element. The metal comes from the metal or base used.

The second part of the salt's name is from the acid used.

Acid used	Type of salt made
Hydrochloric	Chloride
Sulfuric	Sulfate
Nitric	Nitrate

Table 6.2 *Examples of salts made from different acids.*

For example, if you react zinc with nitric acid the salt made will be zinc nitrate:

zinc + nitric acid → zinc nitrate + hydrogen

3 Name the salt made when calcium reacts with hydrochloric acid.

4 Write the word equation for the reaction between:

a) tin and sulfuric acid

b) aluminium and hydrochloric acid.

Reactions between acids and carbonates

Metal carbonates such as calcium carbonate are compounds that are bases – they neutralise acids. Some carbonates, such as sodium carbonate, are soluble and form alkaline solutions. Carbonates react with acids to produce a salt, water and carbon dioxide (a gas). For example, the word equation for the reaction between calcium carbonate and hydrochloric acid is:

calcium carbonate + hydrochloric acid → calcium chloride + water + carbon dioxide

The carbon dioxide is released as bubbles.

Key terms

correlation: relationship (link) between variables where one increases or decreases as the other increases.

reactivity: how likely it is that a substance will undergo a chemical reaction.

6.11 *Marble is a rock that contains calcium carbonate. It reacts with hydrochloric acid to produce bubbles of carbon dioxide.*

5 Indigestion is caused by too much stomach acid.

It can be stopped by taking indigestion tablets that contain magnesium carbonate.

a) Write a word equation to show the reaction that happens between indigestion tablets and stomach acid (hydrochloric acid).

b) Explain why they help stop indigestion.

6 Marble statues can be damaged over time because of acid rain. Suggest why this happens.

6.12

Testing for carbon dioxide

You can collect the gas made when acid reacts with a carbonate and test it to show it is carbon dioxide.

When carbon dioxide is bubbled through limewater, the limewater turns from clear to cloudy. This happens because limewater is a solution of calcium hydroxide. The carbon dioxide reacts with the calcium hydroxide to make insoluble calcium carbonate.

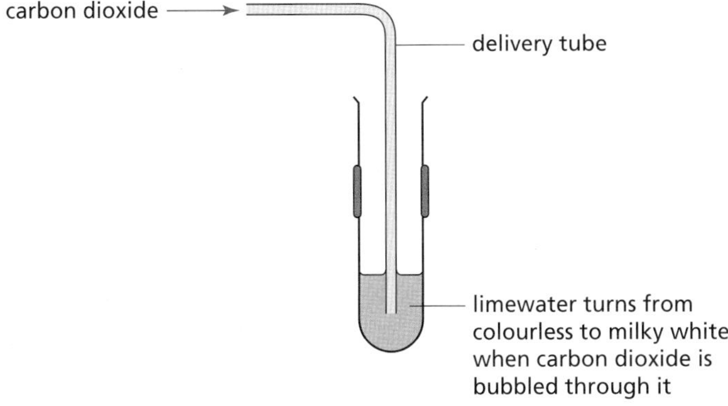
carbon dioxide
delivery tube
limewater turns from colourless to milky white when carbon dioxide is bubbled through it

***6.13** How to test for carbon dioxide.*

7 Write word equations for:

a) The reaction between zinc carbonate and sulfuric acid.

b) What happens when carbon dioxide is bubbled through limewater.

Activity 6.3: Which rock?

Rocks can be used to build many structures such as buildings, sidewalks and statues.

It is important to choose rocks that will last a long time and not be damaged by acid rain, but which rock is best?

A1 You have pieces of the following rocks: marble, granite, sandstone, chalk and limestone. Write down a scientific question that you want to answer.

A2 Plan an investigation method to answer your question. Make sure you include the variable you will change, the variable you will measure and any variables you will keep the same.

A3 State at least *one* way to control risks.

Key facts:

- ✔ Metals and bases react with acids to produce salts. Hydrochloric acid produces chloride salts and sulfuric acid produces sulfate salts.
- ✔ Most metals react with acids to produce hydrogen gas.
- ✔ To test if a gas is hydrogen you use a lighted splint. If a 'pop' noise is heard, hydrogen is present.
- ✔ Carbonates react with acids to produce carbon dioxide.
- ✔ To test if a gas is carbon dioxide, you bubble the gas through limewater. If the limewater goes cloudy, carbon dioxide is present.

Check your skills progress:

I can ask a scientific question and plan an investigation to test this idea.

I can identify important variables, choose which variables to change, control and measure.

I can identify correlations in results.

Reactions with oxygen

Learning outcomes

- To describe and write word equations for the reaction of metals and non-metals with oxygen
- To discuss how ideas can be tested by carrying out investigations and collecting evidence

Starting point

You should know that...	You should be able to...
A reaction with oxygen is called an oxidation reaction	Plan investigations to test ideas
No atoms are made or lost during a chemical reaction; they are just rearranged	Make conclusions from data

Testing theories

Scientists used to think that when a substance burns it loses a substance called phlogiston, which is seen as a flame. Ideas like this are called a **theory.**

Scientists collect evidence to see if their theory is correct.

Charcoal is nearly 100% carbon. When charcoal is burned, its mass decreases. Scientists explained that this happens because the phlogiston leaves the carbon and escapes into the air. This evidence was used to support the phlogiston theory.

6.15 *When carbon is burned its mass decreases.*

However, there are other observations it cannot explain. When magnesium is heated it gives out a bright flame but its mass *increases*. This evidence is not supported by the phlogiston theory.

Key term

theory: idea or set of ideas that explains an observation.

6.14 *Do burning substances lose phlogiston?*

The reaction of carbon and oxygen

We now know that the phlogiston theory is not correct. We know this because of particle theory, which is a newer theory that can explain all the observations. Particle theory explains that everything is made up of atoms. We can use this to explain why the mass of carbon decreases when it is burned and why the mass of other substances, like magnesium, increases.

When carbon is heated in air it reacts with oxygen to form a new compound, carbon dioxide. Energy is transferred as heat and light. This reaction is called **combustion.**

It can be shown as a word equation, and as a particle diagram.

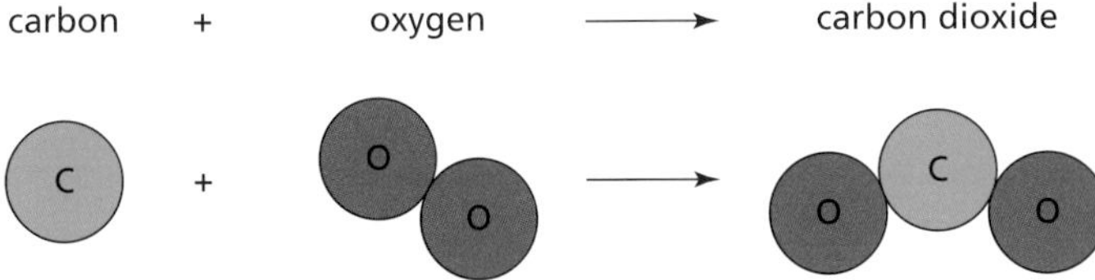

6.17

6.16 *When magnesium is burned its mass increases.*

The mass of the carbon decreases because its atoms are joining with oxygen atoms to form carbon dioxide, which is a gas. The carbon dioxide gas escapes during the reaction and mixes with the air.

The mass of the carbon decreases. But, you can see from the particle diagram that the number of atoms at the start of the reaction is the same as the number of atoms at the end. If you could weigh all the reactants and all the products you would find that they have the same mass.

1. Explain how a new theory (particle theory) helped change ideas about what happens when substances burn.
2. Plan a way of collecting evidence to prove that carbon dioxide is made when carbon burns.
3. When you burn a candle, substances in the wax react with oxygen in the air to produce the products carbon dioxide and water vapour.

   Explain why a candle gets smaller as it burns.

Key terms

combustion: chemical reaction between a substance and oxygen, which transfers energy as heat and light.

Climate change

6.18 *Climate change is linked to an increase in flooding.*

The petrol and diesel we use in cars and lorries is a mixture of compounds called hydrocarbons. These compounds are molecules that contain carbon and hydrogen atoms. When these fuels burn in engines, carbon dioxide and water vapour are produced and mix with the air.

Both these gases are known as greenhouse gases. They trap heat energy, keeping the Earth warm. However, because in the past 100–200 years we have been burning a lot more fuels, the amount of greenhouse gases in our atmosphere is increasing, leading to an increase in the Earth's average temperature. This warming is having an effect on our climate and is linked to increased flooding in some regions.

The reaction of metals and oxygen

When metals are heated in air the metal reacts with oxygen in the air to form an oxide. This is an oxidation reaction.

For example, copper is a shiny orange coloured metal. When it is heated in oxygen it forms a dull, black coating. This is a new compound called copper oxide.

The word equation for this reaction is:

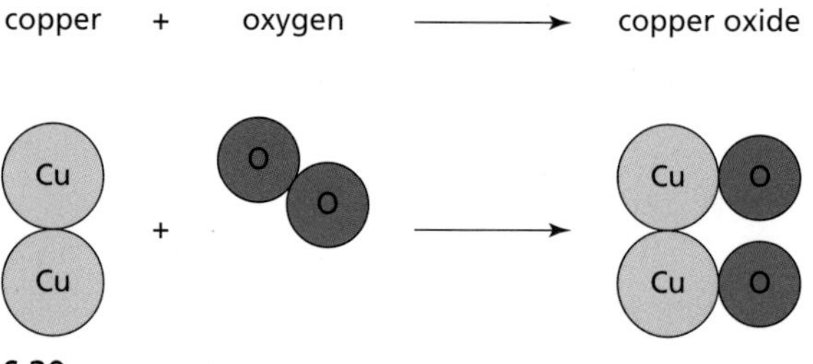

6.20

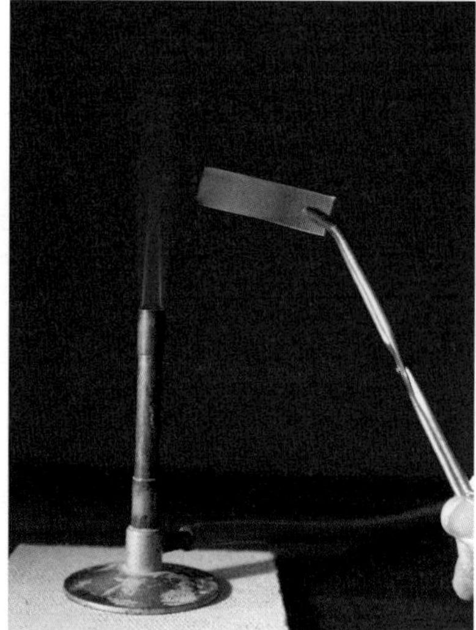

6.19 *When copper is heated in air, black copper oxide is formed on the surface.*

You can measure the mass of the copper before heating, then again afterwards.

You will find that the mass of the copper has increased.

This is because oxygen atoms from the air have reacted with the copper to form copper oxide.

copper + oxygen → copper oxide

10 g 0.5 g 10.5 g

Activity 6.4: Heating magnesium

Elin and Yuki wanted to find out what would happen to the mass of magnesium when it was heated in air.

This is the method they followed:

A Measure the mass of a crucible and lid.

B Place a strip of magnesium into the crucible and measure the mass again.

C Put on your eye protection and place the crucible on the tripod over a Bunsen burner. Heat the crucible.

D Every 30 seconds use tongs to lift the lid slightly off the crucible and place it back down. Do not look at the magnesium directly.

E When the magnesium stops glowing (about 5-10 minutes), turn off the Bunsen burner and let the equipment cool down.

F Measure the mass of the crucible and contents again.

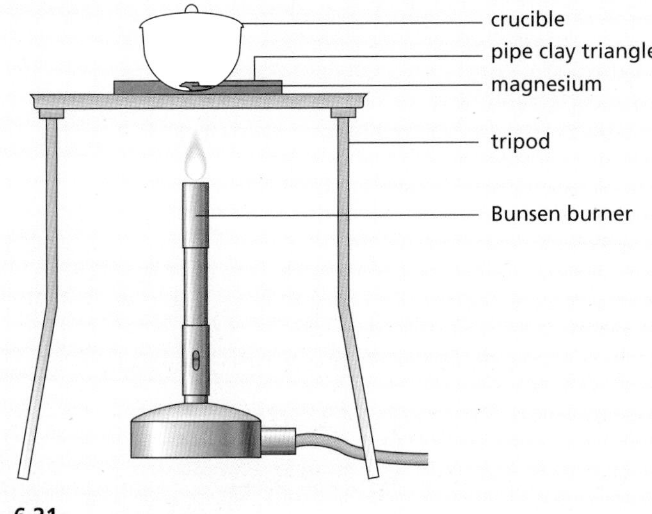

6.21

These are their results.

Mass of crucible = 23.4 g

Mass of crucible + magnesium before heating = 24.6 g

Mass of crucible + magnesium after heating = 25.4 g

A1 Write the word equation for the reaction.

A2 Suggest why a lid was used, and why the lid was opened when the magnesium was being heated.

A3 Calculate the change in mass of the magnesium. Give a scientific explanation for this.

A4 Describe how the risks were controlled.

Key facts:

✔ Oxidation is a reaction with oxygen to form an oxide compound.

✔ Combustion, or burning, is an oxidation reaction.

✔ Some substances react with oxygen and form a solid product, while others react with oxygen and form a product which is a gas.

Check your skills progress:

I can describe how ideas can be tested by carrying out investigations and collecting evidence.

End of chapter review

Quick questions

1. What salt is formed when calcium reacts with sulfuric acid?

   a sulfur calcate **b** sulfuric calcium **c** calcium sulfuric **d** calcium sulfate **[1]**

2. Iron is added to a solution of copper chloride. Copper and iron chloride are formed.

   Write the word equation for this reaction. **[2]**

3. What *two* reactants can be used to make the salt magnesium chloride?

   a magnesium **b** sulfuric acid **c** hydrochloric acid **d** oxygen **[2]**

4. Write the word equation for the reaction when zinc is heated in oxygen. **[1]**

5. What gas is formed when acids react with metals?

   a oxygen **b** nitrogen **c** hydrogen **d** carbon dioxide **[1]**

6. Green copper carbonate powder was heated. The powder started bubbling and it turned black.

   (a) State *two* signs that a chemical reaction took place. **[2]**

   (b) One of the products was a gas. How can you tell? **[1]**

7. A piece of magnesium with mass 2.4 g was heated in air to form magnesium oxide with a mass of 4.0 g. Calculate the mass of oxygen that reacted with the magnesium. **[1]**

8. Explain why combustion is an example of an oxidation reaction. **[1]**

Connect your understanding

9. A carbonate was added to an acid to make the salt potassium chloride.

   Write the word equation for this reaction. **[2]**

10. Deep added an unknown compound (X) to some hydrochloric acid.

    Bubbles were formed.

    (a) Describe how he could find out if the gas was hydrogen or carbon dioxide. **[4]**

    (b) He discovers that the gas is carbon dioxide. What does this show about compound X? **[1]**

    (c) He cannot name compound X. Explain why. **[1]**

11. A teacher measured the mass of a piece of iron wool (fine strands of iron) as 3.6 g. She used tongs to hold the iron in a Bunsen burner flame. She noticed it turned darker and dull in colour. She let it cool and then measured the mass again. It was now 3.9 g.

    (a) Describe what happened to the mass of the iron when it was heated. **[1]**

    (b) Use the particle model to explain why this happened. **[2]**

12. When paper is burned, its mass decreases. Explain why. **[3]**

Challenge question

13. Use the formulas below to draw a particle diagram to show the reaction between calcium and sulfuric acid.

    Sulfuric acid: H_2SO_4

    Calcium sulfate: $CaSO_4$ **[5]**

End of stage review

1. Jordi makes a pot of fresh coffee in his kitchen.

People in other rooms can soon smell the coffee.

(a) What process causes the smell to spread? **[1]**

(b) Write words to fill in the gaps in the sentences to explain why this happens. **[2]**

Coffee particles move from the kitchen, where they are at a ____________________ concentration to other rooms where they are at a ____________________ concentration.

(c) Coffee contains the compound caffeine.

Caffeine has the chemical formula $C_8H_{10}N_4O_2$

(i) What is a compound? **[2]**

(ii) Name the four elements found in caffeine. **[4]**

2. Rashin investigates the best way of preventing an iron nail rusting.

She covers three iron nails with different materials and leaves one nail uncovered.

She leaves the nails in test tubes of water for a week.

The table shows her results.

Material used to cover nail	How rusty the iron nail is after a week (0 = no rust, 5 = very rusty)
paint	1
grease	3
zinc	0
none	5

(a) What conclusion can Rashin make from her results? **[2]**

(b) Suggest why she left one nail uncovered. **[1]**

(c) Rashin reads on a website that the more acidic rain there is, the faster iron will rust.

Write down a plan that she can use to investigate this idea. **[3]**

(d) Write down *two* safety precautions Rashin should take to reduce the risks to herself. **[2]**

3. Students in a class make rockets using empty plastic soda bottles.

(a) One group decides to build a water rocket that uses air pressure.

They use a bicycle pump to pump air into the bottle.

1. Why can't you see an object round a corner?
2. Explain how the shadow of the hand is produced.
3. What would happen to the size of the shadow if you had a bigger hand?
4. Predict what would happen to the size of the shadow if the distance between the light and the hand is increased.

Activity 7.1: Investigating shadows

You will need two pencils of different lengths, a torch and some white paper.

A1 With a partner, find a way to make both pencils produce identical shadows on the paper.

A2 Draw diagrams to show how the shadows are formed.

Include in your drawings the light source, the pencil and the paths of some light rays from the light source to the paper.

Analysing data

Some students used a toy animal to make shadows on a wall. They put a bright light 1 m from the wall. They put the toy at different distances from the wall and measured the height of the toy's shadow. They plotted a graph of their results.

Graphs help us to see the pattern in results. This graph shows that there is a **correlation** between the height of the shadow and the distance of the object from the wall. The height of the shadow increases as the distance of the toy from its shadow increases.

Drawing a **line of best fit** through the points makes the pattern clear. One point does not quite fit the pattern. Odd results like this are called **anomalous**. Perhaps somebody made a mistake when they made the measurement, so it is always a good idea to repeat your measurements.

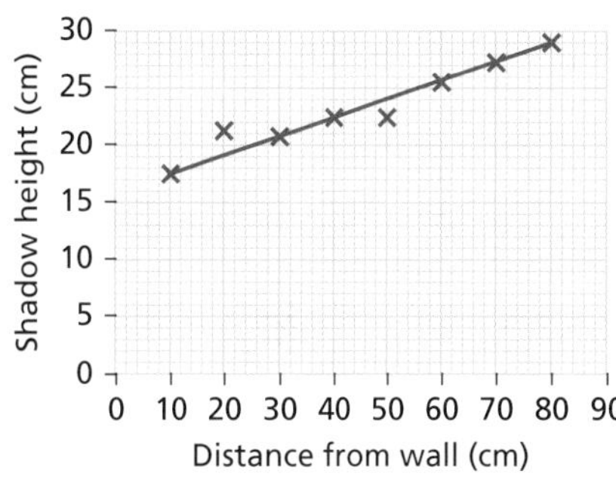

7.2 *Results from experiment.*

Key terms

anomalous result: result that does not follow the same pattern as other measurements.

correlation: relationship (link) between variables where one increases or decreases as the other increases.

line of best fit: a straight or curved line drawn through the middle of a set of points to show the pattern of data points.

Pinhole camera

A 'pinhole camera' is a box with a hole in one end. Some light goes through the hole into the box. This light forms an image on the other end of the box.

Because light travels in straight lines, only light rays from the top of the balloon can reach the bottom of the screen. Only light rays from the bottom of the balloon can reach the top of the screen. The image is upside down.

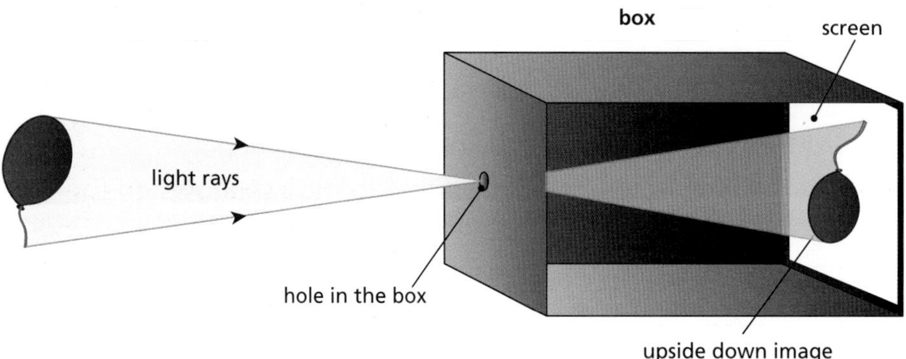

7.3 *How a pinhole camera forms an image.*

The image is very faint. The bigger the hole, the brighter the image. This is because more light rays from the object enter the box. However the image is not as sharp. This is because light from one point on the object can reach more than one point on the screen.

Activity 7.2: Investigating the pinhole camera

Pinhole cameras are easy to make. If you point the hole towards a fairly bright light source and look through the camera you will see the image of the light source on the paper.

7.4 *A simple pinhole camera has no lens.*

A1 Plan how you could investigate the factors that affect the size of the image of a candle flame.

- What will you change and what will you measure?
- Are there any variables that should stay at a fixed value?

A2 Draw a diagram to show how the image of the candle flame is formed. Label the diagram to show the measurements you will take.

A3 Write a prediction. What do you think you will find?

A4 Carry out the experiment and record your measurements.

A5 Draw a graph of your results, with image height on the vertical axis. Describe the relationship between the image height and the variable you investigated.

A6 What do you think will happen to the image if there is more than one hole in the pinhole camera? Explain your prediction before testing it.

Applying ideas: shadows from the Sun

Shadows from the Sun change length during the day. A shadow is shortest at midday, when the Sun is high in the sky.

A shadow's *position* also changes.

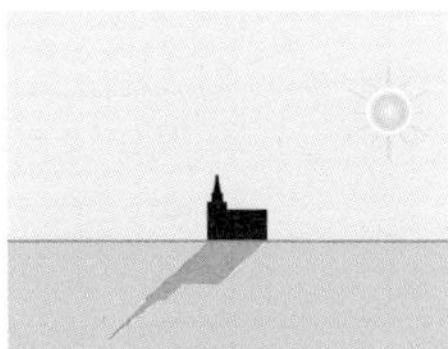

7.6 *At different times of day the shadow will be at a different place.*

7.5 *A shadow from the Sun can be much larger than life.*

5. At what time of day is your shadow the longest?
6. Describe how the length of your shadow changes from morning to night.
7. Explain why the direction of a shadow changes during the day.

Using light and shadows

In India and the Middle East traditional 'mashrabiya' windows give patterned shade and protection from the hot summer sun. Modern architects use a computer programme to plan the effect of shadows before a building is built. The programme uses the position of the Sun at different times of day and the link between an object's height and shadow length to help decide where to plant trees around the building. Preventing sunlight falling on the building keeps it cool and helps reduce air-conditioning costs.

7.7 *Mashrabiya window.*

Applying ideas: eclipses

When the Moon comes directly between the Earth and the Sun there is a **solar eclipse**. The Moon blocks out the light from the Sun.

Key term

solar eclipse: when the Sun looks like it is completely or partly covered with a dark circle.

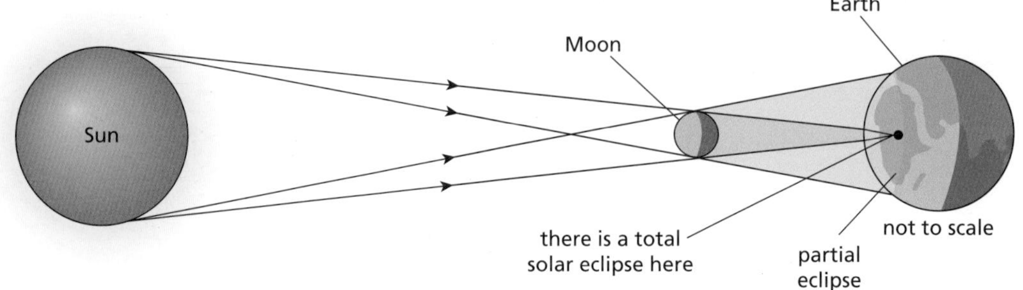

7.8 *An eclipse of the Sun. No sunlight reaches part of the Earth because the Moon blocks the light.*

8 Not all parts of the Earth see a solar eclipse when the Moon is directly between the Earth and the Sun. Why is this? (Hint: Look at the diagram. Which parts of the Earth are in sunlight? What parts of the Earth are in shadow?)

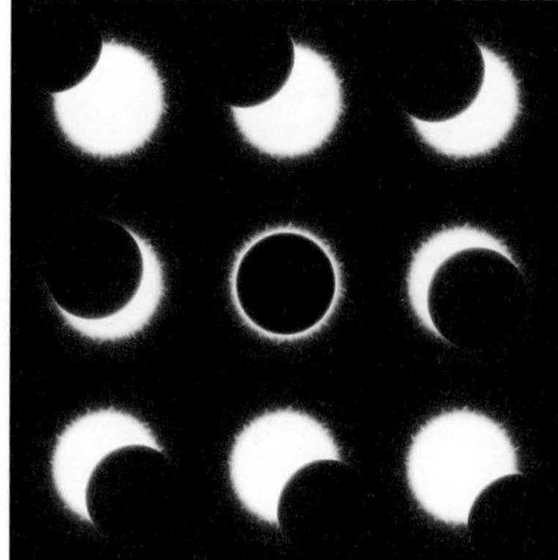

7.9 *An eclipse of the Sun.*

Key facts:

- ✔ A light source gives out light.
- ✔ We see things because light travels from light sources to our eyes or from light sources to objects and then to our eyes.
- ✔ Light travels in straight lines.
- ✔ We draw the path of light with straight lines called light rays.
- ✔ A shadow forms when an opaque object blocks light.

Check your skills progress:

I can describe how to show that light travels in straight lines.

I can draw a ray diagram to explain why a pinhole camera image is upside down.

I can write a plan and choose which variables to change, control and measure.

I can spot a data point that does not fit the pattern.

Reflection

Learning outcomes

- To describe reflection at a plane surface
- To state and use the law of reflection
- To make a periscope

Starting point

You should know that...	You should be able to...
Light travels in straight lines	Do experiments using light rays
A light ray is a straight line which shows the direction of light	Draw diagrams to show how light travels

How light reflects

If you look in a mirror you see yourself. This is called an image. How is this image formed?

Activity 7.3: Investigating how light reflects

Set up the equipment as in figure 7.10.

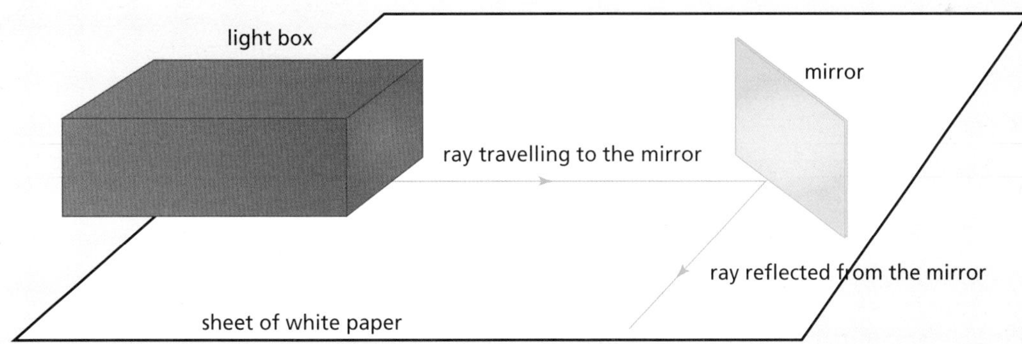

7.10 *Investigating how light reflects.*

A Shine a ray of light from the ray box to the **plane mirror**.

B Use a pencil to draw in the path of the ray towards the mirror (the **incident ray**) and after it reflects (the **reflected ray**). Add an arrow to each ray to show the direction in which it travels.

C Draw a dotted line at 90° to the mirror at the point where the ray hits the mirror (see figure 7.11). This line is called the **normal**.

D Measure the **angle of incidence** and the **angle of reflection** (see figure 7.11).

A1 What do you notice about the angles of incidence and reflection?

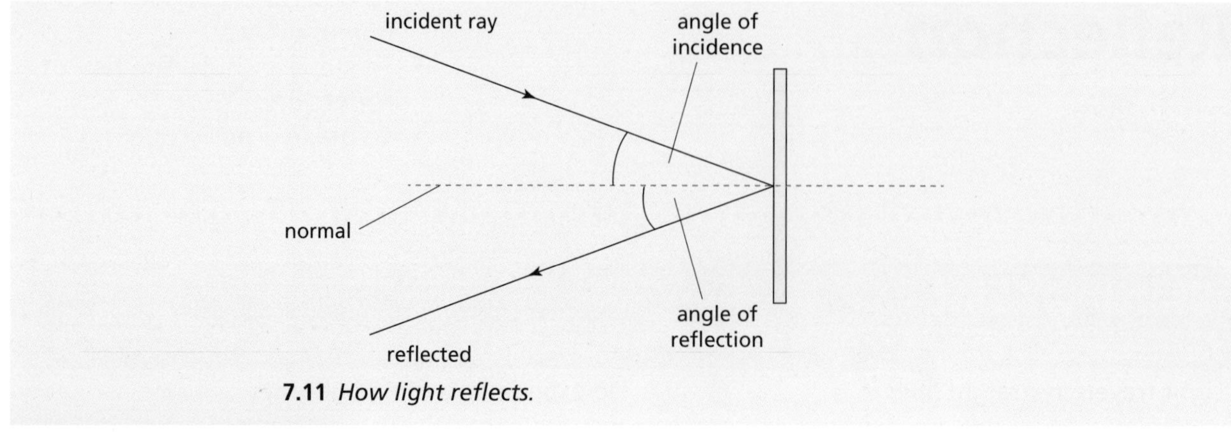

7.11 *How light reflects.*

Law of reflection

The law of reflection is:

angle of incidence = angle of reflection

Remember that both angles must be measured between the ray and the normal.

This law is always true for all shapes of mirror and all angles of incidence.

Some people think that light always reflects through an angle of 90°. Is that correct?

Draw a diagram to explain your answer.

How are images formed in mirrors?

The light from an object is reflected when it reaches the mirror. When this light enters your eye, your brain thinks the light has travelled in a straight line. So, it thinks the light has come from somewhere behind the mirror. This is where the image is formed. The image is always as far behind the mirror as the object is in front.

Look carefully at figure 7.12. You need to be able to draw ray diagrams like these.

Key terms

angle of incidence: this is the angle between the incident ray and the normal.

angle of reflection: this is the angle between the reflected ray and the normal.

incident ray: this ray shows the light travelling towards the mirror.

normal: this is a line drawn at 90° to the mirror at the point where rays hit the mirror.

plane mirror: plane means flat, so a plane mirror is a flat mirror.

reflected ray: this shows the light travelling away from the mirror after it has been reflected.

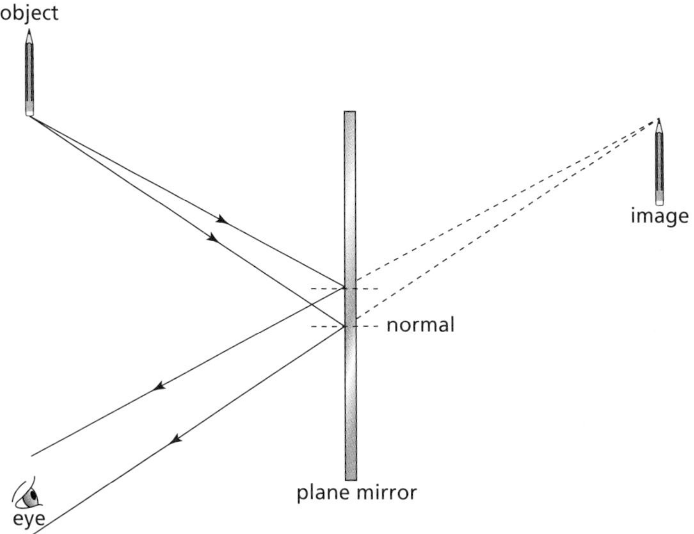

7.12 *How images are formed in mirrors.*

Use figure 7.12 to answer these questions.

2 If you stand 50 cm in front of a plane mirror, how far behind the mirror will your image seem to be? How do you know?

3 What does the letter F look like in a mirror?

Periscopes

Periscopes let us see places we could not normally see. For example, submarines use periscopes so that they can see what is on the surface of the sea even when they are underwater. They use two mirrors.

Figure 7.13 shows a periscope being used to look over a wall.

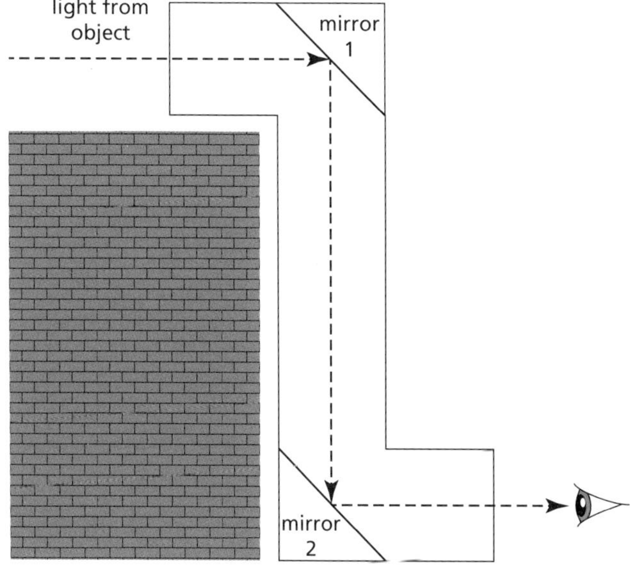

7.13 *A periscope.*

7.14 *Using a periscope.*

Activity 7.4: Making a periscope

Make a simple periscope using a cardboard tube or small box (e.g. a drink carton) and plastic mirrors.

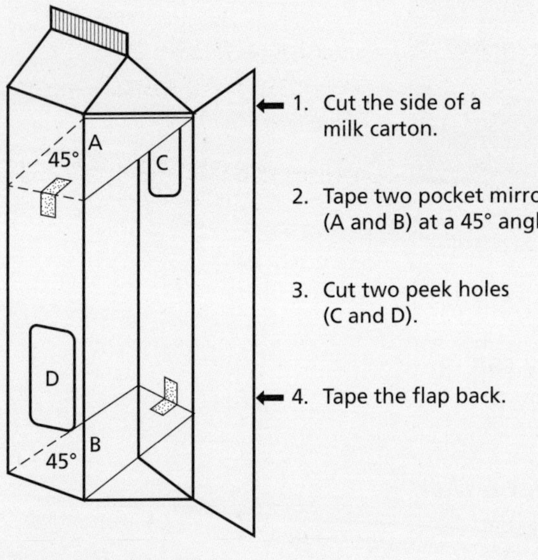
45°
A
C
D
45°
B
1. Cut the side of a milk carton.
2. Tape two pocket mirrors (A and B) at a 45° angle.
3. Cut two peek holes (C and D).
4. Tape the flap back.

7.15 *How to make a periscope.*

Mirrors and solar power

We can collect solar energy and store it using solar panels and batteries. Sometimes mirrors are also used to help concentrate the light from the Sun. This means that more solar power can be produced. Many hot countries generate a lot of their electricity this way. People who live in remote areas can generate their own electricity using solar power too.

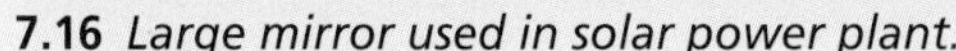

7.16 *Large mirror used in solar power plant.*

7.17 *Many smaller mirrors used in solar power plant.*

Key facts:

- ✔ The angle of incidence = the angle of reflection.
- ✔ The image in a mirror seems to be as far behind the mirror as the object is in front of it.
- ✔ A periscope uses two mirrors to reflect light.

Check your skills progress:

I can do an experiment to show the law of reflection.

I can use diagrams to explain where the image of an object seen in a mirror seems to be.

I can make a periscope and describe how it works.

Chapter 7 . Topic 3

Refraction

Learning outcomes

- To describe refraction
- To give examples of when refraction happens
- To do experiments to investigate how light refracts
- To draw accurately light refracting when travelling between water and air and between glass and air

Starting point

You should know that...	You should be able to...
Light travels in straight lines	Do experiments using light rays
When we draw light rays we add arrows to show the direction of the ray	Draw accurate diagrams to show how light travels
The normal is a line at 90° to a surface drawn at the point where a ray meets that surface	Use a protractor to measure angles accurately
Your brain sees an image where rays entering your eye seem to have come from	

What is refraction?

Why does water look shallower from above than it really is? Why does a straight pencil look bent or broken when part of it is under water?

7.18 *One effect of refraction.*

These things happen because of **refraction**. Light travels through transparent materials such as water and air. If a ray of light travels from one substance (for example, air) into another (for example, water) it often changes direction. This is called refraction.

Key term

refraction: the bending of light when it enters a different medium.

Activity 7.5: Investigating refraction

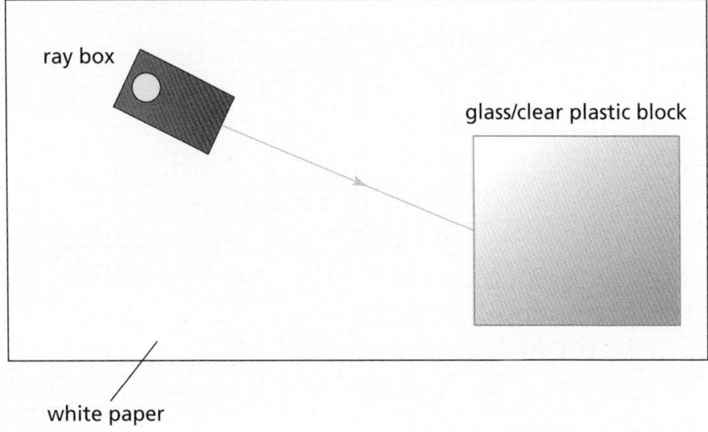

7.19 *Investigating refraction.*

A Set up the equipment as shown in figure 7.19 in a darkened room.

B Draw around the glass block.

C Shine a ray of light into the block.

D Draw the path of the ray from the ray box to the block and after it leaves the block.

E Remove the block. Draw the path of the ray through the block using a straight line to join the places where it enters and leaves.

F Replace the glass block in the same position as before. Move the ray box so the light enters the block at a different angle of incidence.

G Repeat for at least five different angles. Include using an angle of incidence of 0° (where the ray enters the block along the normal).

A1 Look at your drawings and at figures 7.20 and 7.21. Choose the best word to complete these sentences:

- **a** Rays that enter the glass along the normal *bend/don't change direction*.
- **b** The ray bends *towards/away from* the normal when it enters the glass.
- **c** The ray bends *towards/away from* the normal when it leaves the glass.
- **d** The ray that leaves the glass is *parallel/not parallel* to the incident ray.
- **e** Unless the ray enters along the normal, the angle of refraction is always *larger/smaller* than the angle of incidence.

7.20 *Refraction of light when the incident ray is along the normal.*

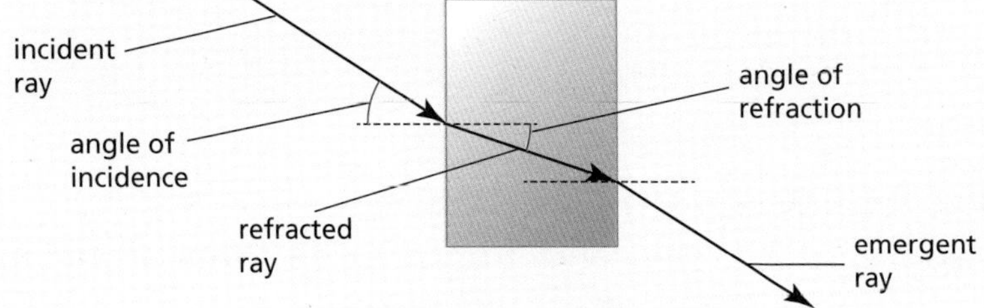

7.21 *Refraction of light for all other angles of incidence.*

Real and apparent depth

Water looks shallower than it really is when viewed from above. Refraction explains this.

Imagine looking down into a stream like the one in figure 7.22. Light from a stone at the bottom travels through the water, into the air and into your eyes. When the light moves from the water into the air, it refracts.

When the refracted light enters your eyes, your brain thinks it has travelled in a straight line and comes from point X (the **apparent depth**). This is where your brain thinks the bottom of the stream (the **real depth**) is.

Key terms

apparent depth: how deep something appears to be.

real depth: how deep something really is.

1. You look at a coin in a glass beaker. Its real depth is 10 cm. Is its apparent depth more than 10 cm or less than 10 cm?

2. Light refracts more when it travels between glass and air than when it travels between water and air.

   You look at a coin 20 cm under water. Its apparent depth is about 15 cm.

   The same coin is under a glass block 20 cm thick. Is its apparent depth more than 15 cm, 15 cm or less than 15 cm?

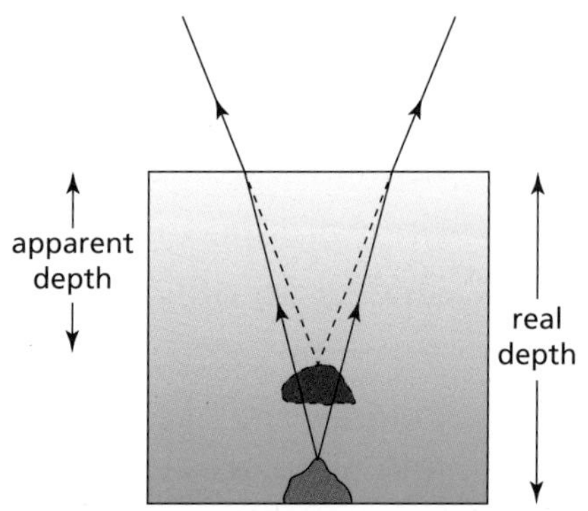

7.22 *Real and apparent depth.*

Refraction can explain many optical illusions.

Applying ideas

Activity 7.6: The appearing coin trick

You will need a coin, a beaker and water

A Put the coin into the beaker

B Look at the coin, then move further away until you can't see the coin. Stay where you are.

C Ask your partner to gradually pour water into the beaker.

D Note when you can see the coin again.

A1 Discuss why you can see the coin again when there is water on top of it.

Activity 7.7: The bending pencil trick

You will need a pencil, ruler or straight piece of wood.

A Half fill a glass with water. Place the pencil in the water so it is half in and half out.

B Look at the pencil from above. Look at the pencil from the side. What do you notice?

A1 Draw a diagram to show what you see from the side. Explain why this has happened.

Mirages

Mirages happen because of refraction. We usually think of mirages as being seen in deserts, but they can happen anywhere. Mirages occur because light refracts and changes direction when it travels from warm air to colder air or from cold air to warmer air.

Look at figure 7.23. Because your brain thinks the light has travelled in a straight line it thinks the blue it can see on the ground is water. It is really an image of the sky. There is no water there at all.

A dry road looks wet for the same reason as you can see in figure 7.24.

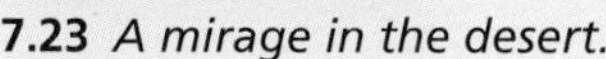

7.23 *A mirage in the desert.*

7.24 *A mirage on a dry road.*

Key facts:

- ✔ Light refracts when it travels from air into glass, clear plastic or water.
- ✔ Refraction makes water look shallower from above than it really is.

Check your skills progress:

I can do experiments to show how light refracts.

I can use a ray diagram to show how light refracts through a rectangular glass block.

Chapter 7 . Topic 4

Coloured light

Learning outcomes

- To explain how white light can be separated into coloured light
- To understand the terms absorption, scattering and dispersion
- To describe examples of how different coloured lights combine
- To investigate how coloured filters work

Starting point

You should know that...	You should be able to...
Light refracts when it travels from air into glass, clear plastic or water	Draw accurate diagrams to show how light travels

Making a spectrum

If you shine white light into a **prism** (see figure 7.25), coloured light comes out at the other side. It looks like a rainbow. Why does this happen? Where do the colours come from?

White light is a mixture of different coloured lights. The prism splits the white light into these different colours. This is **dispersion**.

In figure 7.25, the light entering the prism is refracted when it goes into the prism and again when it leaves. White light is a mixture of all colours. As white light passes through the prism, different colours of light are refracted by different amounts. Figure 7.25 shows that violet light is refracted more than red light.

The colours in the light spectrum are:

Red Orange Yellow Green Blue Indigo Violet

You can remember the order of the colours from '**R**inse **o**ut **y**our **g**reasy **b**ottles **i**n **v**inegar'! You can make up your own saying to help you remember the colours too.

There are other ways of showing that white light is a mixture of the colours of the spectrum.

If you look at white light through a **diffraction grating**, you can also see a spectrum.

Key terms

dispersion: the splitting of white light into a spectrum of colours.

prism: transparent object that refracts light.

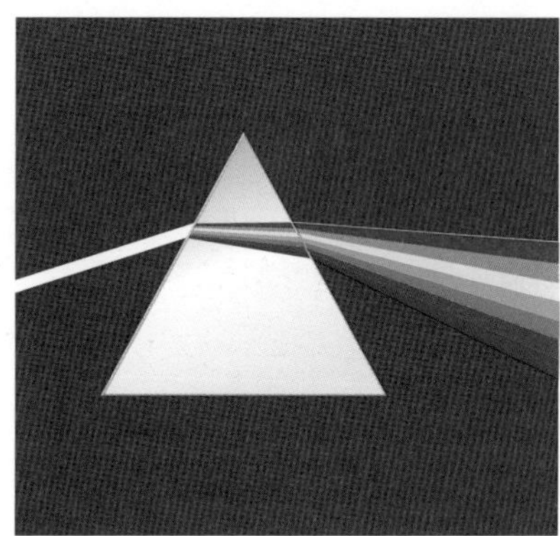

7.25 *The spectrum formed by a prism.*

Key term

diffraction grating: transparent piece of glass or plastic which has many lines drawn onto it. Light can pass through the spaces between the lines.

A diffraction grating is a transparent piece of glass or plastic. It has many lines drawn onto it. Light can pass through the spaces between the lines. There are usually about 3000 lines per centimetre. The closer together the lines are, the clearer the spectrum is. If the lines are too far apart, you won't see a spectrum at all.

7.26 *A diffraction grating.*

Activity 7.8: Making a Newton's disc

A Cut a circle of card. Divide it into seven equal sections. Colour each section as shown in figure 7.27.

B Make a hole through the centre of your disc and put a pencil through the hole.

C Spin the disc as fast as you can. What do you see?

A1 What does this experiment show?

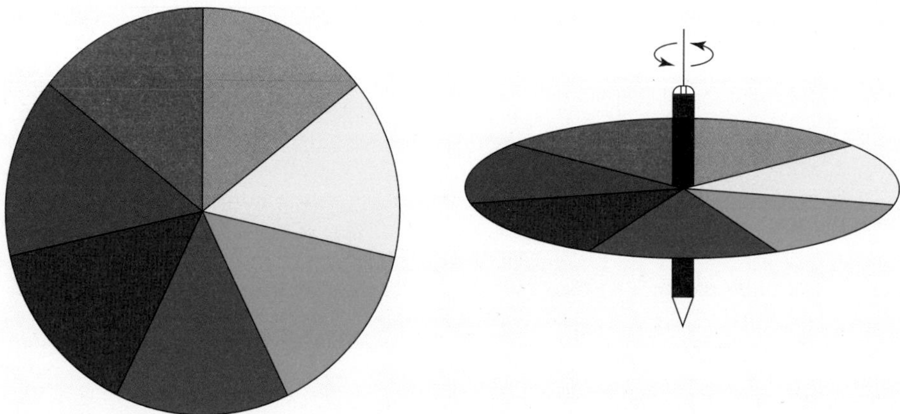

7.27 *Newton's disc.*

Combining colours

Look at figure 7.28. Why does the shirt on the right look blue and the shirt on the left look red? It is because the blue shirt only reflects blue light and the red shirt only reflects red light.

7.28 *Coloured tops.*

Primary and secondary colours of light

There are three **primary colours** of light:

Red **Blue** **Green**

Mixing these colours of light together will make all other colours of light.

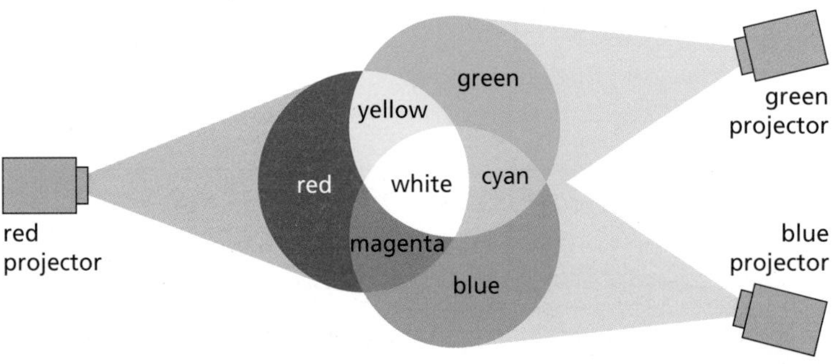

7.29 *Primary and secondary colours.*

Yellow, magenta and cyan are called **secondary colours.**

Objects appear black when they do not reflect any light at all. We say that a black object absorbs all the light that shines on it. The effect is called **absorption**.

You can make many different colours by mixing different amounts of each primary colour. This is how you get coloured images on TVs and computer screens.

Key terms

absorption: the way in which an object takes in the energy reaching its surface.

primary colours: red, blue and green. Mixing these colours of light together will make all other colours of light.

secondary colours: yellow, magenta and cyan.

Colour filters and coloured lighting

A colour **filter** will only let light of its own colour pass through it. The filter absorbs all other colours of light.

For example, a red filter will only allow red light to pass through it.

Figure 7.30 shows what happens if you use different colours of light with different filters.

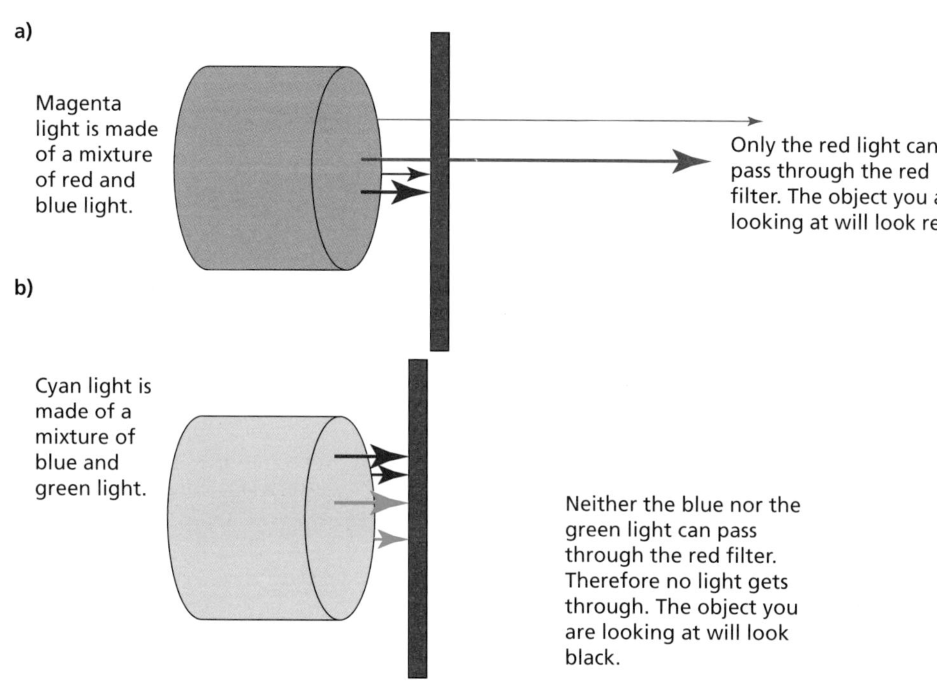
a)
Magenta light is made of a mixture of red and blue light.
Only the red light can pass through the red filter. The object you are looking at will look red.
b)
Cyan light is made of a mixture of blue and green light.
Neither the blue nor the green light can pass through the red filter. Therefore no light gets through. The object you are looking at will look black.

7.30 *What happens when you use a) magenta and b) cyan light with a red filter.*

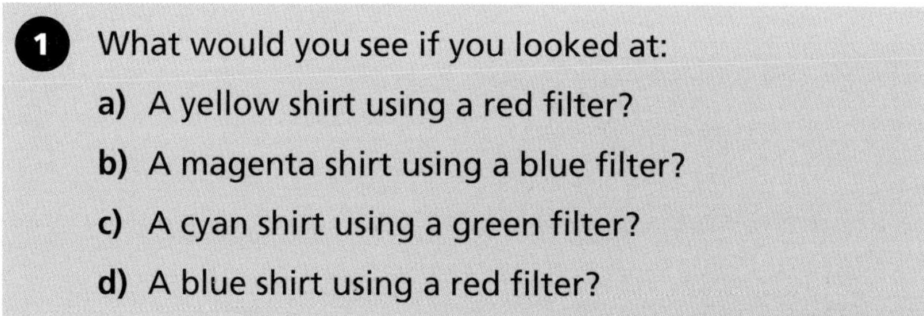

1 What would you see if you looked at:

a) A yellow shirt using a red filter?

b) A magenta shirt using a blue filter?

c) A cyan shirt using a green filter?

d) A blue shirt using a red filter?

You can use colour filters to make different coloured lights. You can use different coloured lights to make objects look a different colour too.

Imagine you are looking at a sheet of white paper. White can reflect all the primary colours. Figure 7.31 shows you what happens when different colour lights are shone on different coloured papers.

Key term

filter: colour filter will only allow light of its own colour to pass through it.

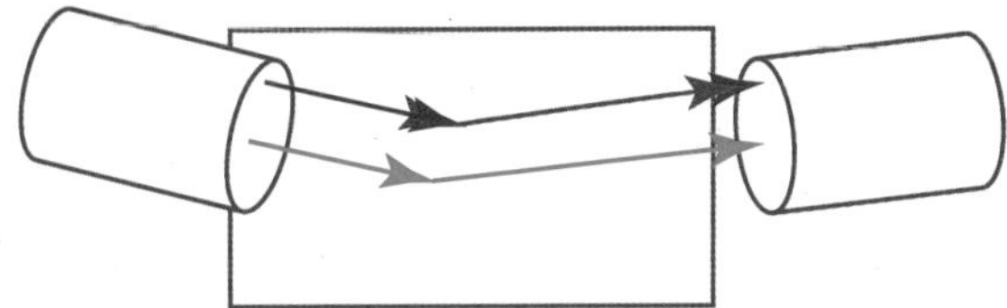

White paper viewed under white light: all three primary colours of light are reflected so the paper looks **white**.

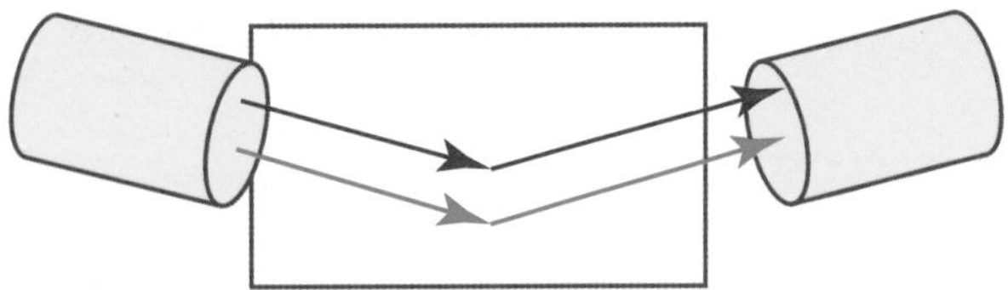

White paper viewed under yellow light: red and green light are reflected so the paper looks **yellow**.

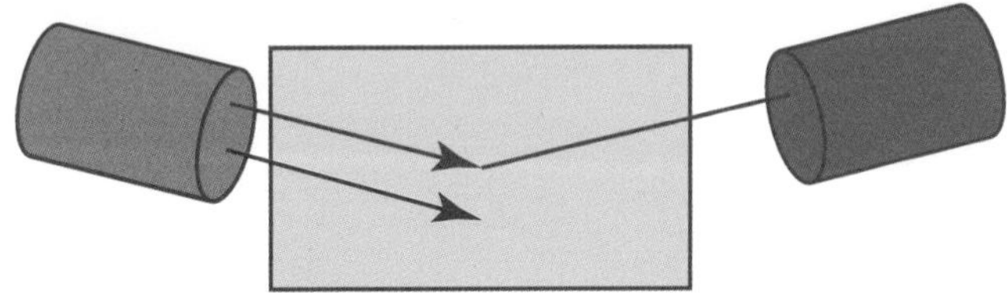

Blue paper viewed under magenta light: because the paper is blue, only blue light will be reflected so the paper looks **blue**.

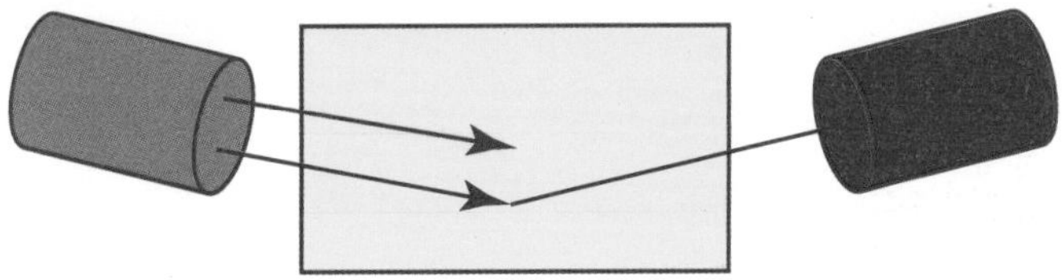

Yellow paper viewed under magenta light: yellow paper only reflects red and green light. So, if you shine magenta light (a mixture of red and blue light) on it only red light will be reflected so the paper looks **red**.

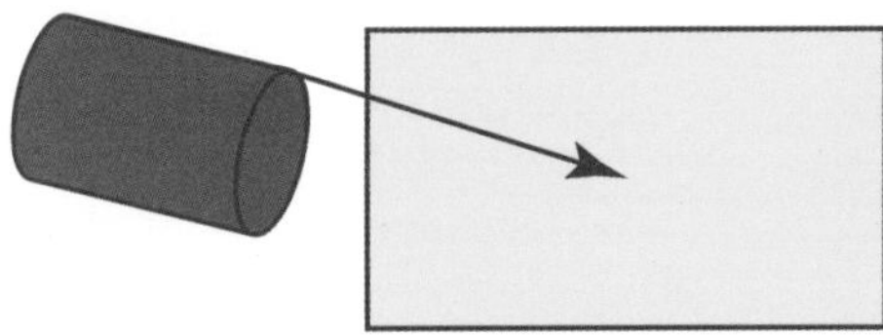

Yellow paper viewed under blue light: yellow cannot reflect blue light so the paper looks **black**.

7.31 *Shining different coloured lights onto different coloured papers.*

2 What would you see if you looked at:

a) A magenta shirt under red light?

b) A red shirt under yellow light?

c) A cyan shirt under blue light?

d) A blue shirt using green light?

Scattering

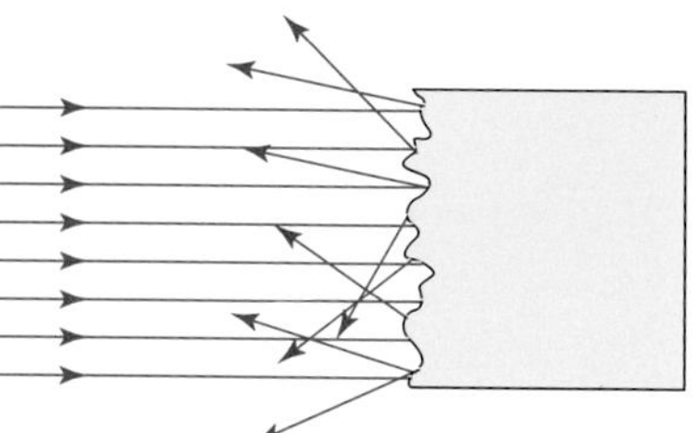

7.32 *Scattering of light.*

Almost everything scatters light. When light hits a rough surface or particles in the atmosphere it reflects. Depending on the shape of the surface, the light will reflect in different directions.

Different colours of light are scattered by different amounts. Violet light is scattered the most and red light is scattered the least.

In the daytime, the sky looks blue because the Sun is high in the sky. Light from the Sun is scattered by the particles in the atmosphere. The angle of **scattering** means that more scattered blue light reaches your eyes. At sunrise and sunset, the Sun is lower in the sky. The angle of scattering means that more scattered red light reaches your eyes.

Key term

scattering: scattering happens when light is reflected from particles and uneven surfaces.

Activity 7.9: Looking through a colour filter

A Look at objects around you through different coloured light filters. What do you notice?

B Use a red pen, a green pen and a blue pen to write on white paper. Look at the writing through a red filter, a green filter and a blue filter.

What do you see?

You could also try writing a secret message that can only be seen using a colour filter.

For example, when viewed through a red filter **WHESELXLLWCOOVMDE** says **WELCOME**.

Activity 7.10: The effects of coloured lights

A Use a ray box (or torch) and colour filters to make different coloured beams of light. Shine these onto different coloured objects and note any colour changes you see. Try to do as many colour combinations as you can.

B You could also try writing a secret message that can only be seen using a coloured light.

For example, when viewed under green light **FHOSELXLLWDOB** says **HELLO**.

Using coloured glass

Coloured glass is often used for decoration.

In the Mosque of Whirling Colours beautiful patterns are seen when light shines through the windows. The colours in the patterns can change slightly near sunrise and sunset.

Many other places of worship use coloured glass windows too.

7.33 *Coloured glass in the Nasir-al-mulk mosque in Iran.*

Key facts:

- ✔ A prism can split white light into a spectrum. This is dispersion.
- ✔ The primary colours of light are red, blue and green.
- ✔ Combining primary colours can make secondary colours and white.
- ✔ Filters only allow certain colours of light to pass through them.
- ✔ Filters absorb other colours of light.
- ✔ Scattering takes place when light falls on a rough/uneven surface or when it hits particles in the air.

Check your skills progress:

- I can do experiments to show how a prism can form a spectrum.
- I can use filters to make different coloured lights and make objects look a different colour.
- I can predict what something will look like under different colour lights and through different coloured filters.

End of chapter review

Quick questions

1. **(a)** Describe how shadows are formed. **[1]**

   (b) How could you make the shadow made by an object smaller? **[1]**

2. What do each of the letters A, B, C, D and E stand for in the diagram below?

   Choose from these words.

angle of incidence	angle of reflection	incident ray	normal	reflected ray

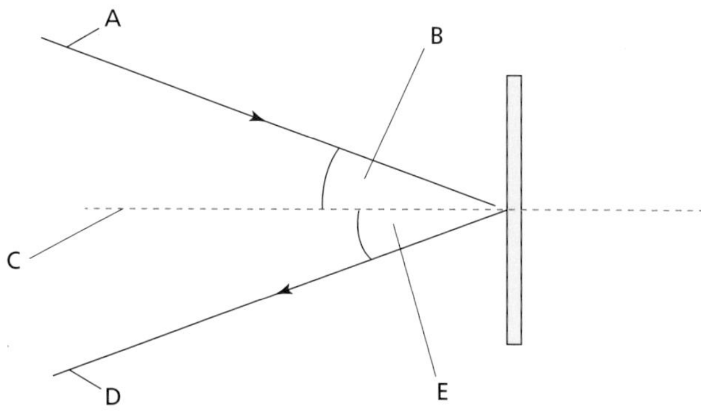

7.34 *Reflection in a plane mirror.* **[5]**

3. Challenge You can sometimes see a rainbow when it is raining. A water droplet can behave like a prism. Suggest how a rainbow is formed. **[2]**

4. Shoppers often take clothes outside to look at them in daylight before deciding whether to buy. Why do they do this? **[2]**

5. What would the writing below say when viewed through a red filter?

   J A R E X E F D D P L U I M Y G J J H B C T **[1]**

Connect your understanding

6. What would you see if you looked at:

   (a) A magenta T-shirt under blue light? **[1]**

   (b) A red T-shirt under cyan light? **[1]**

   (c) A cyan T-shirt under green light? **[1]**

   (d) A magenta T- shirt under green light? **[1]**

   (e) A blue T-shirt under cyan light? **[1]**

7. **(a)** Draw a ray diagram to show how a pinhole camera forms an image of a tree. **[3]**

(b) How does the image compare to the real tree? **[2]**

8. Challenge What would the writing below say when viewed through a yellow filter?

J T H E A N O N D W K Y M O Q U C **[1]**

9. **(a)** Copy and complete figure 7.35 to show where the fish appears to be. **[1]**

(b) Explain why the fish appears to be in the position you have drawn it. **[3]**

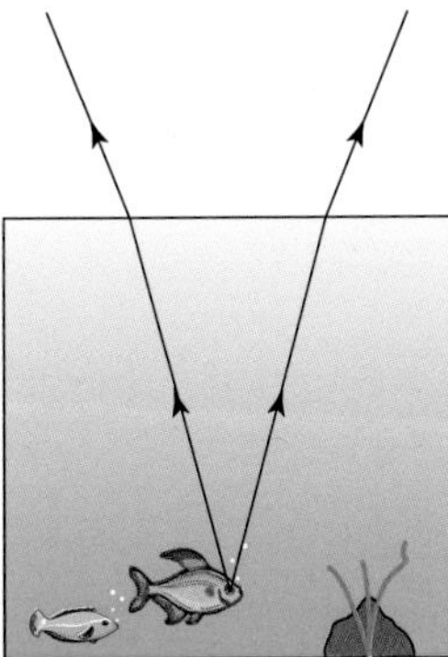

7.35

10. Challenge Look at the arrangement of filters below. The first filter is blue and the second is magenta.

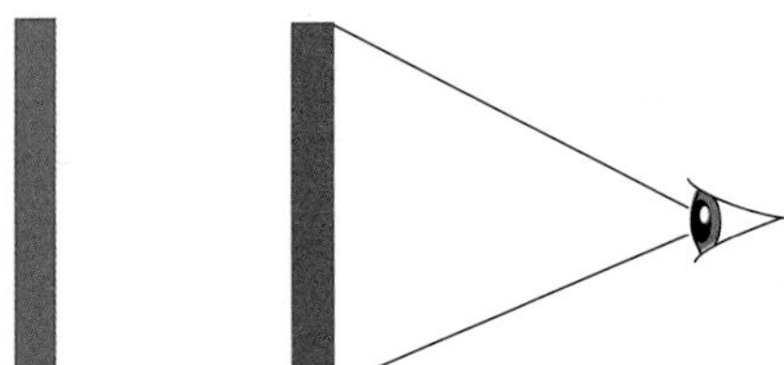

7.36

(a) What colour would you see if the source gave out:

- red light
- cyan light
- yellow light
- white light
- magenta light? **[5]**

(b) Would it make any difference if the filters were the other way around? **[1]**

11. What would the tops in figure 7.37 look like when viewed with:

(a) green light **[1]**

(b) magenta light **[1]**

(c) cyan light **[1]**

(d) yellow light? **[1]**

7.37

12. If you shine white light into a prism it produces a spectrum. This is one piece of evidence to support the idea that white light is made from a mixture of different coloured lights.

Give *two* more pieces of evidence that show white light is made from a mixture of colours. **[2]**

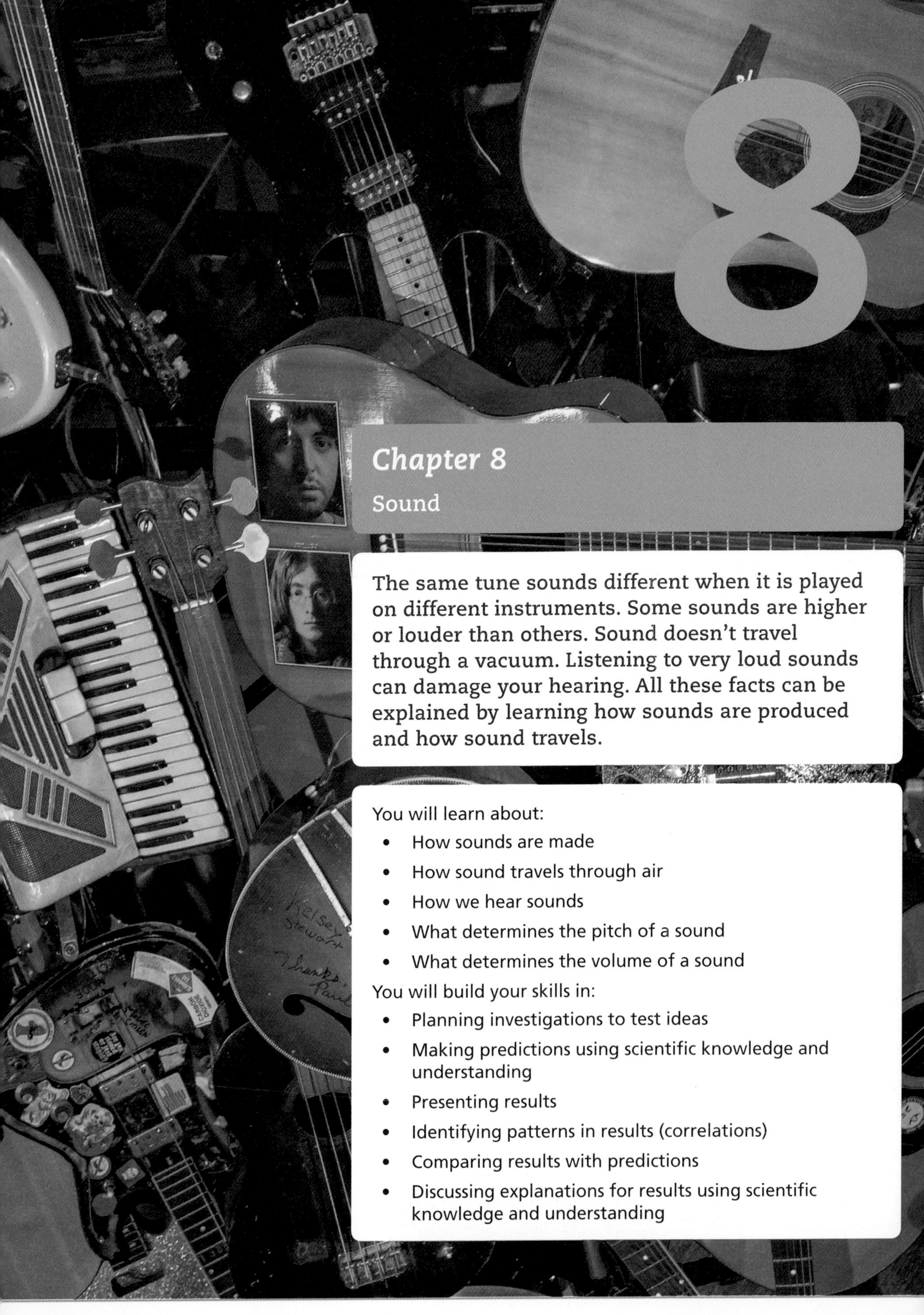

Chapter 8

Sound

The same tune sounds different when it is played on different instruments. Some sounds are higher or louder than others. Sound doesn't travel through a vacuum. Listening to very loud sounds can damage your hearing. All these facts can be explained by learning how sounds are produced and how sound travels.

You will learn about:

- How sounds are made
- How sound travels through air
- How we hear sounds
- What determines the pitch of a sound
- What determines the volume of a sound

You will build your skills in:

- Planning investigations to test ideas
- Making predictions using scientific knowledge and understanding
- Presenting results
- Identifying patterns in results (correlations)
- Comparing results with predictions
- Discussing explanations for results using scientific knowledge and understanding

Chapter 8 . Topic 1

How sounds are made

Learning outcomes

- To describe how vibrations can produce sound
- To investigate how sounds are made
- To describe how sounds require a medium through which to travel

Starting point

You should know that...	You should be able to...
Sound is a form of energy	Suggest ideas that may be tested
Matter is made up of particles	Make careful observations
Particles are arranged differently in solids, liquids and gases	Present results in the form of tables, bar charts or line graphs

How are sounds made?

Clapping your hands makes a sound. Pulling on a piece of elastic and then letting it go makes a sound too. So does blowing into or across a tube. How are these sounds made?

Sounds are made by **vibrations**. When an object vibrates, the air around it vibrates. These vibrations in the air travel to our ears. We detect this as sound.

Key term

vibration: when something moves back and forwards many times, we say it vibrates.

How many different ways can you think of to make air vibrate and create sound?

Making air vibrate

Look at figure 8.1. When you pull the elastic as in A, it moves to the position shown in B. It then springs back and moves to position C. It will vibrate backwards and forwards between B and C and then returns to its original position shown in D.

As the elastic moves, it makes the air particles surrounding it move too. These air particles vibrate backwards and forwards with the elastic.

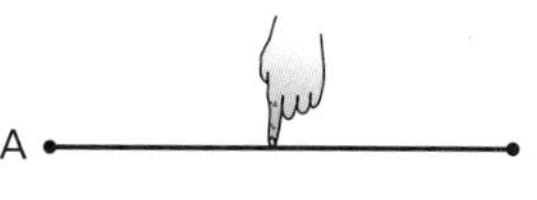

D

8.1 *Using 'strings' to make sound.*

Sound waves

Whenever we make sounds, some air particles are pushed closer together at first and then spring apart again. Others are pulled further apart and then spring back until they are closer together again. The air particles are made to vibrate. This results in a sound wave being created.

Figure 8.2 shows what happens to the air particles close to an object when it creates a sound.

Normal position of air particles.

Position of air particles when a sound is produced. Some are pushed closer together (a **compression**), others move further apart (a **rarefaction**).

The positions of the air particles changes as the particles vibrate. This creates a sound wave which travels through the air.

8.2 *How sound travels through air.*

Key terms

compression: region where particles in a longitudinal wave are closer together.

longitudinal wave: in a longitudinal wave, the wave travels in the same direction as the vibrations that produced it.

rarefaction: region where particles in a longitudinal wave are further apart.

This type of wave is called a **longitudinal** wave. In a longitudinal wave, the wave travels in the same direction as the vibrations that produced it.

Figure 8.3 shows a longitudinal wave moving along a slinky spring. The wave has been started by pushing the left hand side of the slinky spring.

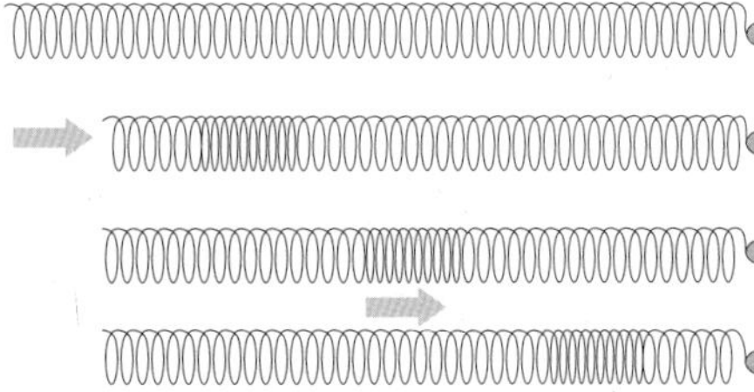

8.3 *Longitudinal wave moves along a slinky spring.*

To create sounds loud enough to hear, you need to get large amounts of air to vibrate.

2 Why do drums have large surfaces?

Sound can travel through all solids, liquids and gases. We call the substance sound travels through the **medium**. The particles of the medium vibrate as the sound travels through it. Sound cannot travel through a **vacuum** (empty space).

8.4 *Sticks can be used to make a sound with a drum.*

3 How are sounds made?

4 How does sound travel through the air?

5 Explain why sound cannot travel through a vacuum.

6 Why does sound travel faster through water than through air? (Hint: think about how the particles are arranged in water and in air)

7 Do you think the speed of sound through steel would be faster or slower than through water? Explain your answer.

Key terms

medium: the substance a sound wave travels through.

vacuum: completely empty space.

Activity 8.1: Music in the classroom

Look around your classroom.

How many musical instruments could you make from the things you can see?

How would you use them?

For each instrument, what has to be made to vibrate to make sounds?

Present your results in a table.

8 We have folds of tissue, called vocal folds or vocal cords in our throats. These vocal folds vibrate when we speak – see figure 8.5.

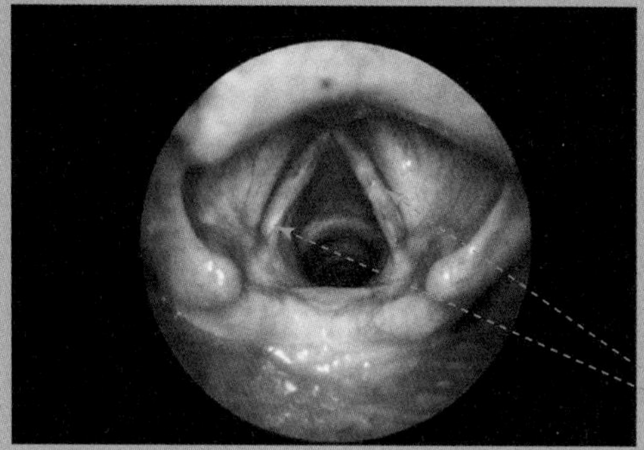

8.5 *Vocal folds inside the throat.*

a) How are they made to vibrate? (Hint: think about how you breathe!)

b) Suggest a reason why people's voices sound different.

Music around the world

Different countries and cultures have developed different types of musical instruments. For example, the balalaika in Russia, pan pipes in South America, bagpipes in Scotland, steel band instruments in the Caribbean and the sitar in South Asia.

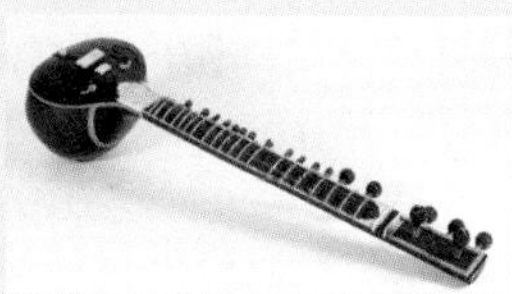

8.6 *Instruments from different parts of the world.*

Key facts:

- ✔ Vibrations produce sound.
- ✔ Sound travels as a longitudinal wave.
- ✔ Sound travels through the air by making the air particles vibrate.

Check your skills progress:

I can make sounds in many different ways.

I can explain how musical instruments make sound.

I can describe how sound travels from its source to our ears.

How we hear sounds

Learning outcomes

- To identify the different parts of a human ear
- To describe how we hear sounds
- To describe how humans hear a limited range of sounds that changes as we get older
- To describe how we can protect our hearing from damage
- To measure the speed of sound

Starting point

You should know that...	You should be able to...
Sounds is made by vibrations	Make sounds in different ways
Sound travels as a longitudinal wave	
Sound travels through air by making the air particles vibrate	

How do our ears detect sound?

You know that sound travels as a longitudinal wave. When a sound is made, it causes the surrounding air particles to vibrate. If these vibrations enter your ears, you hear the sound.

Look at figure 8.7. It shows not only the outside of an ear, but also the parts that are inside the head.

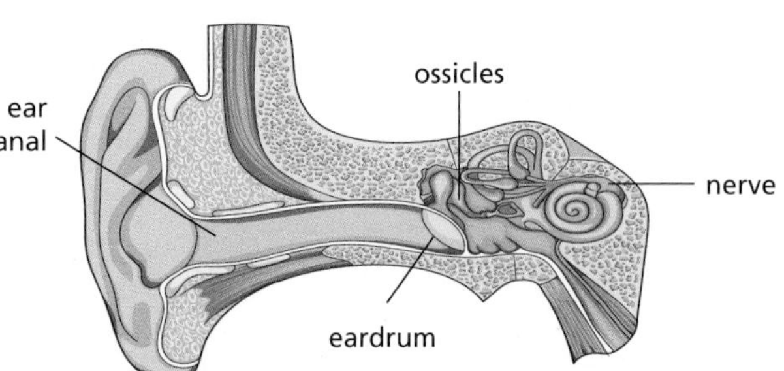

8.7 *The human ear.*

When the vibrating air enters your ear, it travels along the ear canal until it reaches the eardrum.

The eardrum is a very thin piece of tissue. When the vibrating air hits the eardrum, it makes the eardrum vibrate. These vibrations are then passed on and cause three very small bones, called ossicles, to vibrate too.

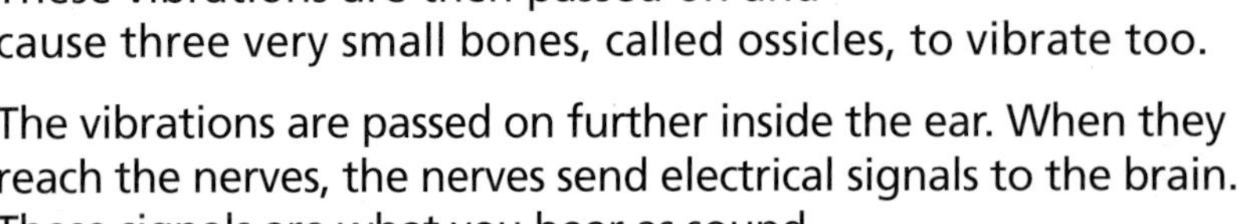

The vibrations are passed on further inside the ear. When they reach the nerves, the nerves send electrical signals to the brain. These signals are what you hear as sound.

1 Why is it difficult to hear if you have a lot of earwax in your ear?

How sound can damage our hearing

The louder a sound is, the harder the air particles hit your eardrum. This is why loud sounds can hurt you and even damage your ears. Sometimes the air particles hit your eardrum so hard that it rips. If this happens you won't be able to hear well in that ear until the eardrum is repaired.

8.8 *Ear protectors help to prevent damage from loud noises.*

People who work in noisy places or use noisy machinery must wear ear protection. This stops the air particles hitting your eardrum too hard.

Headphones focus all the sound into your ear canal. Listening to loud music through them can damage your hearing permanently.

2 A person has an eardrum with a hole in it. Explain why they cannot hear well.

How quickly do you hear a sound after it has been made?

Once a sound wave reaches your eardrum, you hear it immediately. But, before the sound reaches your eardrum it has to travel from its source to your ear. This can take some time.

8.9 *Using headphones too much can damage your hearing.*

The speed of sound through air is about 340 m/s. This is quite fast, but it is slow enough for there to be a noticeable delay between a sound being made and you hearing it.

During a thunderstorm, you see the lightning immediately but it can be several seconds before the sound that the lightning makes (thunder) reaches you. You can use this to work out how far away the thunderstorm is.

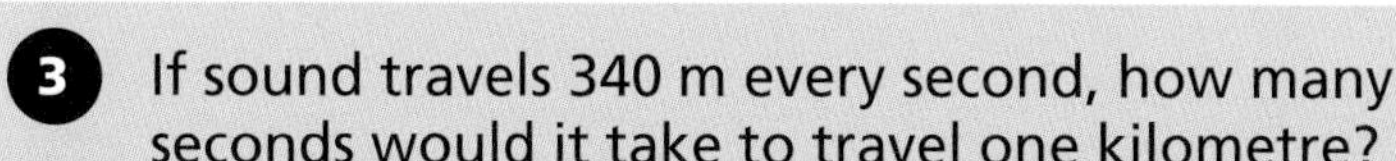

3 If sound travels 340 m every second, how many seconds would it take to travel one kilometre?

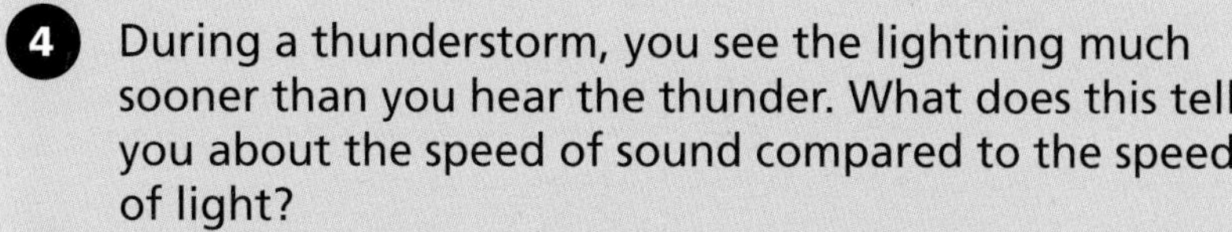

4 During a thunderstorm, you see the lightning much sooner than you hear the thunder. What does this tell you about the speed of sound compared to the speed of light?

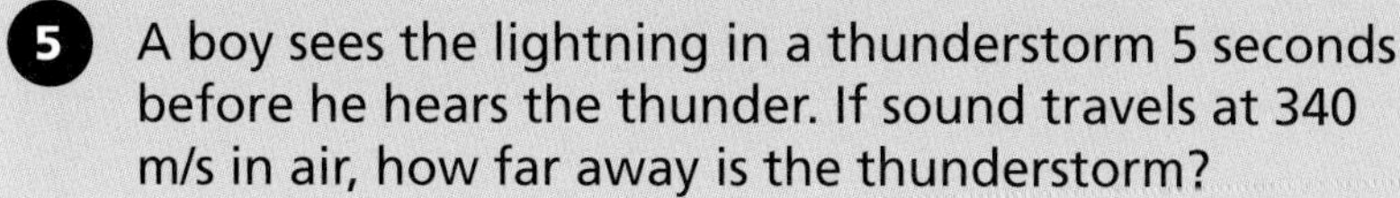

5 A boy sees the lightning in a thunderstorm 5 seconds before he hears the thunder. If sound travels at 340 m/s in air, how far away is the thunderstorm?

Echoes

Echoes happen when sound is reflected.

Sometimes you can hear the echo of your own voice. This can happen if you shout when you are near a large hard object like a cliff or a wall. The sound waves you make reflect from the wall back to you.

8.10 *How echoes are made.*

6 Why do you not hear an echo if you shout when you are *very* close to the cliff?

7 A girl shouts at a cliff. She hears the echo 3 seconds later. If sound travels at 340m/s, how far away from the cliff is she?

Activity 8.2: Measuring the speed of sound

You can measure the speed of sound through air by timing how long it takes for a sound to travel a known distance.

This method needs two people, and needs to be done in a large, open area. You need something that will make a loud, clear noise and is big enough to see from a distance of about 400 m (for example, a drum or two large blocks of wood). You also need a stopwatch and a tape measure.

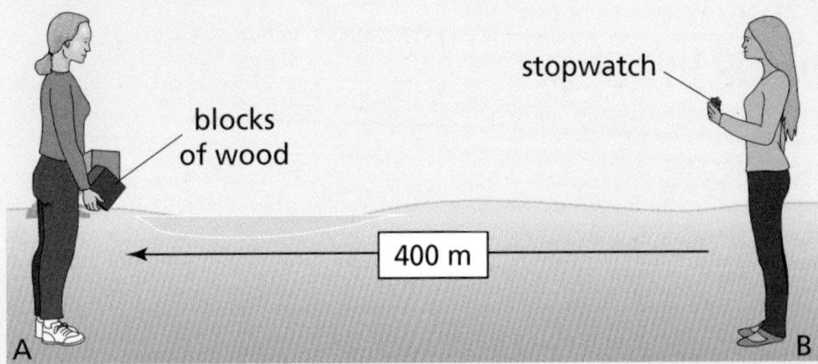

8.11 *Measuring the speed of sound in air.*

A One person (A) stands at one end of the open space with the equipment to make the sound.

B The second person (B) takes the stopwatch and stands as far away as possible. A and B must be able to see each other clearly.

C A makes the loudest sound they can.

D B starts the stopwatch as soon as they see A making the sound.

E B stops the stopwatch when they hear the sound.

F Measure the distance between A and B.

Use the equation speed = $\frac{\text{distance}}{\text{time}}$ to work out the speed of sound.

A1 Why should you repeat the experiment several times, and find the mean of the results?

A2 How could you use echoes to measure the speed of sound through air? Plan an experiment to do this.

Activity 8.3: Comparing ears in different animals

Different animals have different types of ears. Some are large, some are small, some point forwards, some point sideways.

A1 Choose *two* animals with very different types of ears. Describe each animal's ears and suggest why they are like this.

Using sonar for underwater exploration

Sonar is a way of using sound (echoes) to calculate distances and depths. It has been used to find shipwrecks. In 2016 more than 40 shipwrecks were found in the Black Sea.

Sonar is also used to make maps of the bottom of the ocean. These maps show how the depth of the sea changes. Underwater mountains and volcanoes have been found in the South China Sea using sonar.

8.12 *Image of underground volcano found in 2016 in the South China Sea.*

Mapping the depth of the oceans makes sea travel safer as there is less chance of underwater rocks damaging boats.

Key facts:

- ✔ Vibrating air particles enter our ear canals and we detect this as sound.
- ✔ Echoes are sound reflections.
- ✔ Sound travels at a speed of about 340 m/s in air.

Check your skills progress:

I can measure the speed of sound in air.

I can identify experimental errors and suggest how to reduce them.

I can design an experiment to measure the speed of sound in air using echoes.

Loudness and pitch

Learning outcomes:

- To identify the different variables we can change and measure for waves
- To relate these variables to the sounds we can make or hear

Starting point

You should know that...	You should be able to...
Sound is produced by vibrations	Measure the speed of sound through air
Sound travels as a longitudinal wave	
Sound travels at about 340 m/s through air	

Why do sounds have different pitches and loudness?

If you pluck a guitar string, it will make a sound. If you shorten the string it will make a different sound. If you pluck it harder it will make a louder sound. Why is this?

It is because their sound waves are different. How loud a sound is depends on the **amplitude** of the sound wave. The greater the amplitude of the sound wave, the louder the sound. The amplitude of a sound depends on how much energy the sound wave has. For a musical instrument, this depends on how hard you hit, blow or pluck the instrument. Amplitude is measured in metres (m).

The **pitch** of a sound depends on the **frequency** of the sound wave. The greater the frequency, the higher the pitch. Sounds made by a flute have a higher frequency (and a higher pitch) than those made by a cello. The more waves you make every second, the higher the frequency is.

Frequency is measured in **hertz** (Hz). One wave every second gives a frequency of 1 Hz.

Waves with a high frequency will have a short **wavelength** because they will be close together. The shorter the wavelength, the higher the sound. The wavelength of a wave means how long one wave is. In a longitudinal wave, one wavelength is the distance between one compression and the next.

Wavelength is measured in metres (m).

wavelength

8.13 *The wavelength of sound.*

Key terms

amplitude: the maximum height of the wave, from the centre to the top or bottom.

frequency: the number of waves per second.

hertz: the unit of frequency. 1 Hz = 1 complete wave every second.

pitch: the pitch of a sound is how high or low it is.

wavelength: the length of one complete wave.

Looking at sound waves on an oscilloscope screen

A microphone converts sound into electrical impulses. When you connect a microphone to a cathode ray oscilloscope, the electrical impulses are displayed as a graph on a screen. The graphs are called traces.

Though sound waves are longitudinal waves, the traces you see on an oscilloscope do not look like longitudinal waves. They look like a different type of wave, called a transverse wave. The high and low points on transverse waves represent the points where particles are closest together and furthest apart on a longitudinal wave.

Figure 8.14 shows some sound waves on an oscilloscope screen.

Each screen shows the same time period, so you can see how many waves have been made in a given time. All the pictures have the same scale.

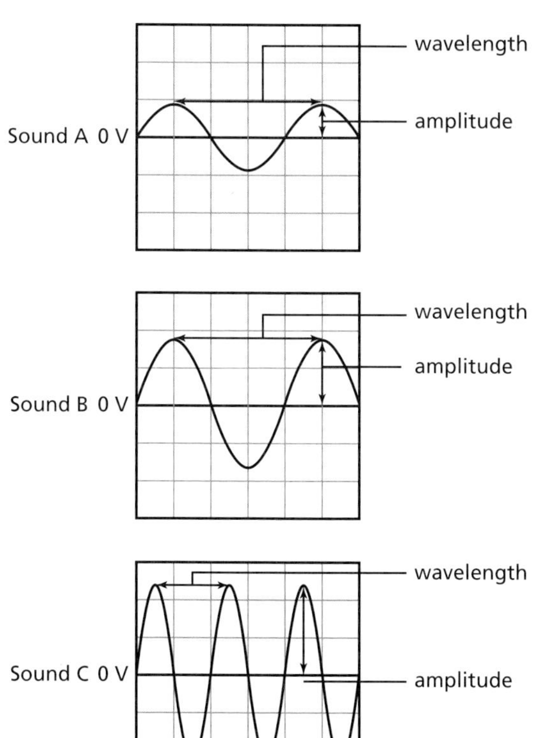

8.14 *Sound waves shown on an oscilloscope screen.*

- Volume
  - Sound A is quieter than sounds B and C. We know this because it has a smaller amplitude.
- Pitch
  - Sound A has the same pitch as sound B. We know this because they have the same wavelength and frequency. The same number of waves have been made in a given time.
  - Sound C has a higher pitch because it has a shorter wavelength and a higher frequency. More waves have been made in the same time.

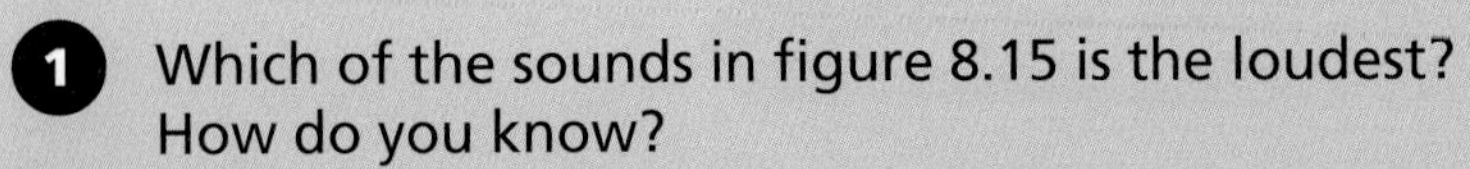

1 Which of the sounds in figure 8.15 is the loudest? How do you know?

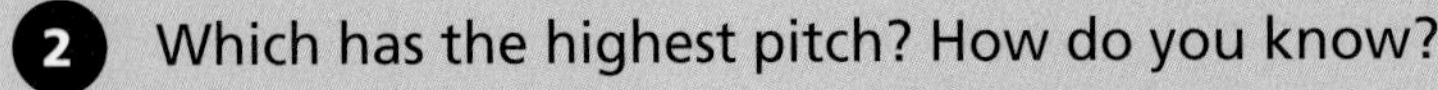

2 Which has the highest pitch? How do you know?

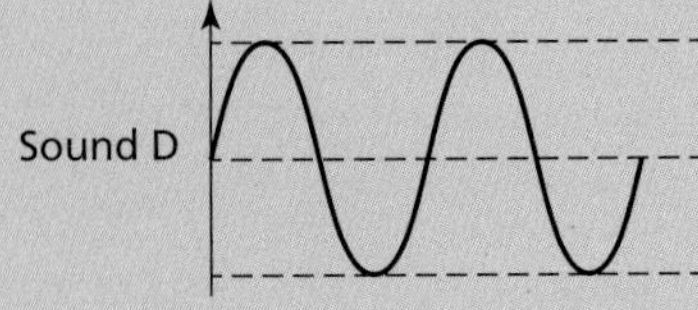

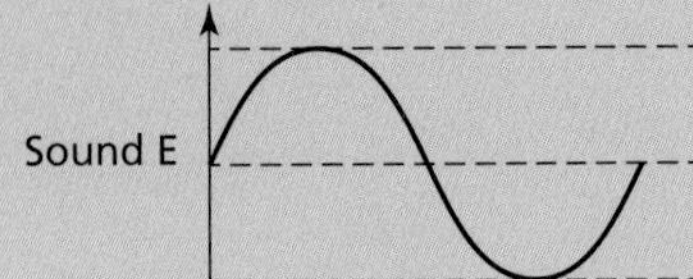

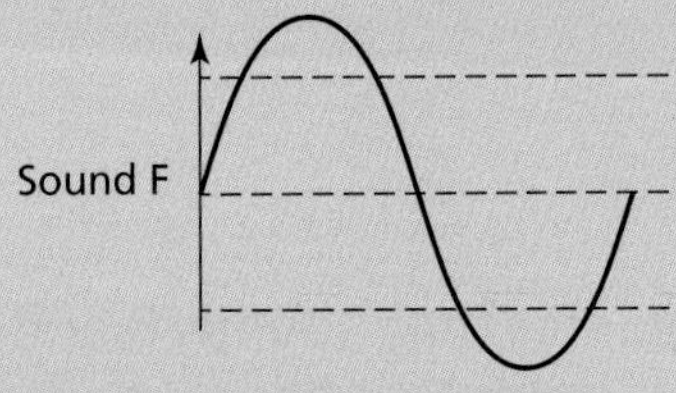

8.15 *Sounds D, E and F.*

Usually, the bigger a musical instrument is, the lower the frequency of the notes it can make. This is because the size of the instrument sets the maximum wavelength of the sound waves it makes. The bigger the instrument, the longer the wavelength. This is why you can change the pitch of the note made by a stringed instrument like a guitar by shortening or lengthening the string.

3 Will shortening a guitar string make the pitch of the note higher or lower? Explain your answer.

Frequency range of human hearing

The human ear can hear frequencies as low as about 20 Hz and as high as about 20 000 Hz (20 kHz).

Anything lower than 20 Hz is called **infrasound**. We cannot hear this but the vibrations are still there.

Elephants can hear sound with frequencies as low as 5 Hz, but they cannot hear frequencies above 10 000 Hz.

Key term

infrasound: sound waves with a frequency too low for humans to hear.

Sounds waves with a frequency higher than 20 000 Hz are called **ultrasound**. We cannot hear these but some animals can.

Dogs have a hearing range of about 40 Hz to 60 000 Hz. Cats can hear even higher frequencies, up to 85 000 Hz. Dolphins can hear a very large range of frequencies, from less than 1 Hz to 200 000 Hz. Bats have a range of about 20 Hz to 120 000 Hz. Dolphins communicate with each other using ultrasound; bats do the same. Bats also judge distances using ultrasound echoes. This helps them to navigate.

Key term

ultrasound: sound waves with a frequency too high for humans to hear.

4 Which of the animals dogs, cats, dolphins and bats:

a) Can hear the lowest frequency?

b) Can hear the highest frequency?

c) Can hear over the biggest range of frequencies?

5 Why can very high-pitched sounds hurt you and possibly damage your hearing? (Hint: think about what happens inside your ears when you hear sounds.)

Activity 8.4: Testing your hearing range

Signal generators can make electrical waves with different frequencies. If you attach a signal generator to a loudspeaker you will get sound waves of different frequencies. The dial on the signal generator will tell you what the frequency is.

Set the signal generator at a very low frequency, too low to hear. Gradually increase the frequency and note when you can first hear a sound. Keep increasing the frequency (and the pitch of the sound) until you can no longer hear it. Note when this happens.

Activity 8.5: Changing the pitch

Get a glass beaker and a pencil. Tap the side of the beaker with a pencil. Listen to the note that this makes.

Now add a little water to the beaker and repeat. Add even more water and repeat again. Keep doing this until you can see a clear pattern to your results.

Activity 8.6: Design a musical instrument

Design a musical instrument made from everyday materials. You must be able to play a simple tune with this instrument, and so it must be able to make at least five different notes.

Using ultrasound in medicine

Ultrasound has many uses. One of its most important uses is in medicine. Ultrasound machines can be used instead of X-rays to diagnose problems in the body. Ultrasound machines are cheaper, much smaller and more portable than X-ray machines.

This means that they can be made available in many more places, even in very remote parts of the world. Studies in Rwanda, Liberia, Ghana and remote parts of the Amazon showed that using ultrasound machines has improved the diagnosis and treatment of illnesses.

Key facts:

- ✔ The pitch of a sound depends on the frequency of the sound wave.
- ✔ The volume of a sound depends on the amplitude of the sound wave.
- ✔ The human hearing range is from 20 Hz to 20 000 Hz.
- ✔ Other animals may have different hearing ranges.
- ✔ Ultrasound means sound waves which are too high for humans to hear.

Check your skills progress:

Understand what waves with different frequencies look like on an oscilloscope screen.

Understand what waves with different amplitudes look like on an oscilloscope screen.

Discuss the risks of listening to sounds which are very loud or very high pitched.

Describe how to use a microphone to make sound waves which can be seen on an oscilloscope.

End of chapter review

Quick questions

1. What does *pitch* mean? **[1]**
2. What feature of a sound wave determines its pitch? **[1]**
3. What does *volume* mean? **[1]**
4. What feature of a sound wave determines how loud it is? **[1]**
5. Copy and complete by filling in the gaps.

   Choose from the words in the list.

ossicles	eardrum	electrical	340

   Sound is made by vibrations. Air particles are made to vibrate. These vibrations travel through the air at a speed of about ____________ m/s. When the vibrations reach your ear they make your ____________ vibrate. This makes the tiny bones called ____________ vibrate. The vibrations are passed on further inside the ear. When they reach the nerve, ____________ signals are sent to the brain. We hear these signals as sound. **[4]**
6. How do you lower the pitch of the note made by a stringed instrument like a guitar? **[1]**
7. Why do large musical instruments usually make deeper sounds than small musical instruments? **[1]**

Connect your understanding

8. **(a)** How can you change the pitch of the sound a drum makes? **[1]**

   (b) Explain your answer. **[3]**
9. How could you use a microphone and an oscilloscope to see what the sound waves made when you speak look like? **[2]**
10. Why do professional musicians, like those who play in orchestras, sometimes have problems with their hearing as they get older? **[2]**
11. Copy the sound wave in figure 8.16 on the right.

    (a) Draw a sound wave that is louder but has the same pitch. **[2]**

    (b) Draw a sound wave that has the same volume but a lower pitch than the first wave. **[2]**

    8.16

    (c) Draw a sound wave that has a higher pitch and a lower volume than the first wave. **[2]**

12. **(a)** How could you use a loudspeaker, a microphone, an oscilloscope and a signal generator to compare how well sound can travel through different materials? Design an experiment to compare wood, glass and steel. **[5]**

(b) State the precautions you would take to make sure your results were accurate. **[2]**

(c) Challenge A singer wants to make one room in their house soundproof. This means that very little sound gets into or out of the room. How would the singer use the results of your experiment to decide which material to use? **[3]**

13. A girl sees the lightning in a thunderstorm 8 seconds before she hears the thunder. If sound travels at 340 m/s in air, how far away is the thunderstorm? **[2]**

14. A boy is standing 400 m away from a cliff. He makes a loud noise. If sound travels at 340 m/s, how long will it be before he hears the echo? Show your working. **[2]**

Challenge questions

15. As you get older, your ears become worse at detecting high-pitched sounds. Suggest why. **[2]**

16. A person makes a sound which starts at a very low pitch. The pitch then becomes higher and higher. At the same time, the sound gets quieter and quieter. Sketch a diagram to show what the sound wave would look like. **[2]**

17. **(a)** How would you use the equipment shown in figure 8.17, to show that sound cannot travel through a vacuum? **[2]**

(b) How does it prove that sound cannot travel through a vacuum? **[1]**

(c) Explain why sound cannot travel through a vacuum. **[1]**

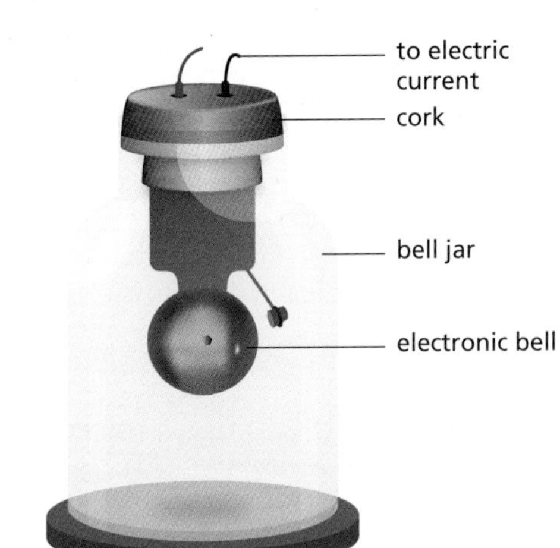

8.17

18. Sound travels faster through water than through air. Suggest a reason for this. **[2]**

9

Chapter 9

Measuring motion

Being able to measure distance and time accurately is important in our world today. In sport, times of 1/100th of a second can make the difference between winning or losing a race. In space science, a spacecraft being a few centimetres out of position can mean it will miss its target and its mission will fail.

You will learn about:

- How distance and time are measured accurately
- How to calculate the speed of a moving object
- How to describe changes in the way an object moves
- How to use graphs to show how an object is moving

You will build your skills in:

- Planning investigations to test ideas
- Making predictions using your understanding of science
- Presenting results in tables and graphs
- Identifying patterns in results
- Comparing results with predictions
- Discussing explanations for results using your understanding of science

Measuring distance and time

Learning outcomes

- To select suitable measuring apparatus
- To measure the distance an object travels accurately
- To measure the time for an object to travel between two places accurately

Starting point

You should know that...	You should be able to...
When things speed up, slow down or change direction there is a cause	Choose appropriate apparatus and use it correctly
Energy is associated with movement	Take accurate measurements
Particles are arranged differently in solids, liquids and gases	Make repeated measurements

When you investigate how an object is moving you need to measure the distance travelled and the time taken. You need to think about how **accurate** your measurement needs to be and the best equipment to use.

Key term

accurate: accurate measurement is where the measurement is very close to the real value.

Choosing the best apparatus to measure distance

Choose the best apparatus from the box to measure each of the following distances.

15 cm ruler marked in millimetres
metre rule marked in millimetres
10 m measuring tape marked in centimetres
100 m measuring tape marked in centimetres

a) How far you can run in 10 s.

b) The height of a table.

c) The length of your thumb.

d) How far you can walk in 2 s.

When you measure small distances, you need to be more accurate than when you are measuring larger distances. For example, you would measure the width of your text book to the nearest millimetre, but the height of a wall to the nearest centimetre.

Activity 9.1: Measuring distance

Measure the following:

- height of your chair
- length of your pencil or pen
- width of your classroom
- length of your thumbnail.

A1 What did you use to measure each distance?

A2 For each measurement, write whether it is accurate to the nearest millimetre or centimetre.

2 Why is it more accurate to measure the width of your desk with a metre rule than a 15 cm one?

Choosing the best apparatus to measure time

9.1 *Different apparatus for measuring time.*

There are lots of ways to measure time (see figure 9.1). If you want to measure how long a lesson lasts, then you can use an ordinary clock or watch. If you need to measure the time to run 20 m you need more accurate equipment.

3 Why is it important to repeat your measurements if you are using a stopwatch to time how long it takes a toy car to travel down a ramp?

Activity 9.2: How long does it take?

Time how long it takes you to do the following:

A Write your name.

B Read a page of this book.

C Count to 200.

A1 How could you have made your measurements more accurate?

Measuring the height of mountains

9.2 *Mount Everest.*

The official height of Mount Everest, 8848 metres above sea level, was calculated in 1955. At that time, we could not measure distances as accurately as we can now.

Now we use satellites which send radio waves to receivers. The receivers are placed at different points on Earth. By knowing the position of the satellite, the speed of the radio waves and the time it takes the radio waves to travel to the receiver, we can calculate the height of the receiver.

There is a receiver near the top of Mount Everest. Signals sent to the receiver mean that we can measure the height of the mountain more accurately than ever before. Recent measurements suggest that Mount Everest is slightly lower than it was in the past. This is likely to be because large earthquakes have affected the ground below it.

Key facts:

- ✔ Rulers and measuring tapes can be used to measure distance.
- ✔ Clocks, watches, stopwatches and light gates connected to electronic timers, data loggers or computers can be used to measure time.
- ✔ Different pieces of apparatus give measurements with different levels of accuracy.

Check your skills progress:

I can choose suitable apparatus to measure distance and time.

I can explain my choice of measuring apparatus.

Speed and average speed

Learning outcomes

- To calculate the average speed of moving objects
- To use scientific knowledge to explain the results of experiments

Starting point

You should know that...	You should be able to...
Some apparatus is more accurate than others	Select appropriate equipment to measure distance and time
Gravity is a force that attracts objects towards the Earth	Measure distance and time accurately
Friction, including air resistance, is a force that slows down moving objects	Calculate the average (mean) of a set of numbers

Measuring speed

When you talk about an object's **speed**, you are describing how far it travels in a given time. Speed is measured in metres per second (m/s). A speed of 10 m/s means the object travels 10 m in 1 s.

1 How far does a car with a speed of 12 m/s travel in:

a) 10 s?

b) 1 minute?

c) 10 minutes?

As you travel to school your speed changes many times. We often measure the **average** speed of the moving object during a journey. This takes account of the object speeding up, slowing down and even stopping during its journey.

$$\text{average speed} = \frac{\text{total distance travelled}}{\text{total time taken}}$$

2 A girl takes 25 s to run 100 m. What is her average speed?

3 A man walks 450 m at an average speed of 3 m/s. How long is he walking for?

Key terms

average: the mean average of a set of numbers is found by:

$$\frac{\text{total of all the numbers added together}}{\text{how many different numbers there are}}.$$

speed: how far something moves in a given time.

Activity 9.3: Measuring your average walking speed

A Measure out a distance of between 5 and 10 metres.

B Use a stopwatch to time how long it takes you to walk this distance.

C Calculate your average speed.

A1 You timed how long it takes to walk a set distance. Describe a different way of doing this experiment.

A2 Why is it better to do this experiment with longer distances and times than shorter distances and times?

In science we often measure speed in metres per second, but in everyday life speed is often measured in different units, for example kilometres per hour (km/h) (see figure 9.3).

9.3 *Speedometer.*

4 Estimate the average speed of the following:

a) an Olympic sprinter running 100 m in m/s

b) a tortoise in cm/s, then convert your answer to m/s

c) a snail in mm/s, then convert your answer to m/s

d) an Olympic marathon runner in km/h. Convert your answer to m/s

e) a cheetah running at full speed in km/h. Convert your answer to m/s.

Using light gates

Light gates are electronic sensors. They can measure time very accurately. In figure 9.4 the timer starts when the front of the card passes through the light gate and stops when the end of the card passes through it. This tells you how long it took the card to travel through the light gate. If the card is 10 cm long, you now know how long it took the car to travel 10 cm.

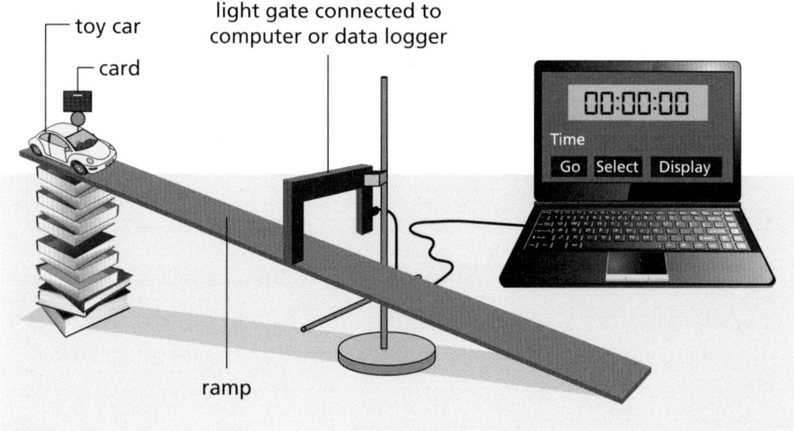

9.4 *Using a light gate to measure time.*

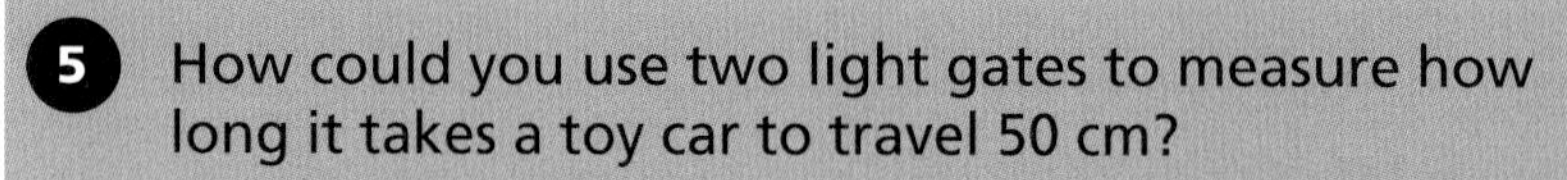

5 How could you use two light gates to measure how long it takes a toy car to travel 50 cm?

Activity 9.4: Measure the average speed of a toy car descending a ramp

Use one of these methods to find the average speed of a toy car as it rolls down a ramp.

Method 1: With a stopwatch

A Time how long it takes a toy car to go from the top to the bottom. Repeat at least three times, then calculate the average time.

B Now calculate the average speed of the car.

Method 2: With a light gate (see figure 9.4)

A Attach a card to the top of a toy car. Set up a light gate halfway along the ramp.

B Release the car at the top of the ramp and record the time taken for the card to travel through the light gate. Repeat at least three times, then calculate the average time.

C Now calculate the average speed of the car.

A1 Why is it important to place the light gate *halfway* along the ramp in method 2?

Anomalous results are results that do not follow the pattern of other results. You are less likely to get anomalous results if you repeat your experiment more than once. Anomalous results can affect your average values if they are very much higher or lower than the rest of your results.

6 Ali timed how long it took a toy car to roll down a ramp. Here are his results.

10 s, 12 s, 14 s, 24 s, 11 s, 13 s, 10 s, 11 s, 12 s, 10 s

a) Which result is anomalous?

b) What is the average time if the anomalous result is included?

c) What is the average time if the anomalous result is not included?

Activity 9.5: Measure the average speed of a toy car descending a ramp (2)

A Plan an experiment to investigate how the average speed of a toy car depends on the steepness of the ramp.

B Design a table to record your results.

C Consider how to display your results as a graph.

A1 What do you expect your experiment to show?

D Check your plan with your teacher before doing your experiment.

A2 Did you get any anomalous results? If so, can you think of a reason for this?

A3 Use your knowledge of gravity to explain your results to the experiment.

Tracking migrating animals

GPS (Global Positioning System) tracking is a new way of tracking the movement of animals. A radio receiver is placed on the animal. The radio receiver picks up signals from special satellites. There is a small computer attached to the receiver and this calculates where the animal is and how it is moving. The data collected is sent to scientists via another satellite. You can now get very small, solar powered GPS receivers – even small enough to attach to a bird. Scientists can use the data to try to discover why the animals are moving and also to discover the reasons for any changes in migration patterns.

9.5 *Migrating birds.*

Key facts:

✔ Average speed = distance travelled / time taken.

✔ Speed is measured in metres per second (m/s).

Check your skills progress:

I can plan an experiment to measure average speed.

I can measure the average speed of a moving object.

I can present results in tables and graphs.

Chapter 9 . Topic 3

Distance–time graphs

Learning outcomes

- To use graphs to show how an object moves
- To understand a distance–time graph for a moving object
- To draw a distance–time graph for a moving object
- To use a distance–time graph to calculate speed

Starting point

You should know that...	You should be able to...
Speed is measured in metres per second (m/s)	Select appropriate equipment to measure distance and time
Average speed $= \dfrac{\text{distance travelled}}{\text{time taken}}$	Draw line graphs with sensible axes, labels and scales
	Find the gradient of a graph

What are distance–time graphs?

A distance–time graph shows if an object is moving and if its speed is constant, increasing or decreasing.

The distance–time graph in figure 9.6 shows how long it takes a runner to run 10 m, 20 m and 30 m.

How can you tell that the runner is moving at a constant speed?

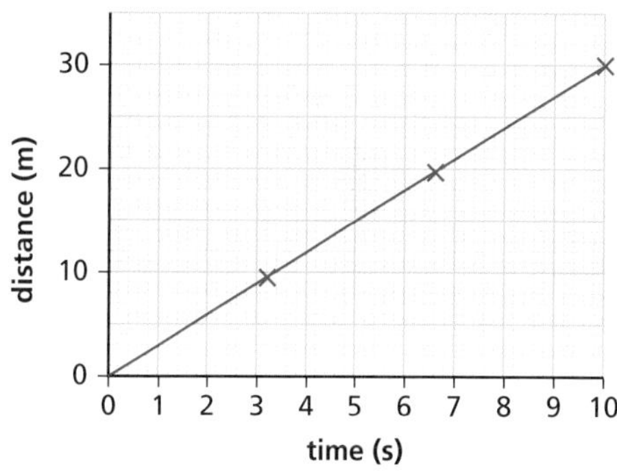

9.6 *Distance–time graph showing a constant speed.*

Calculating speed

To work out the average speed of the runner in figure 9.6 after 10 s:

$$\text{average speed} = \frac{\text{distance travelled}}{\text{time taken}} = \frac{30\ \text{m}}{10\ \text{s}}$$

This is the same as working out the **gradient** (slope) of the graph.

$$\text{gradient} = \frac{\text{distance travelled}}{\text{time taken}}$$

For the runner in figure 9.6:

$$\text{gradient} = \frac{30\ \text{m}}{10\ \text{s}}$$

Key term

gradient: the gradient of a graph tells you how steep the line is.

Figure 9.7 shows a distance–time graph for a faster runner.

2 What is the speed of the runner in figure 9.7?

3 Graphs 9.6 and 9.7 both have the same axes. Which graph has the steepest slope – the one for the faster runner or the one for the slower runner?

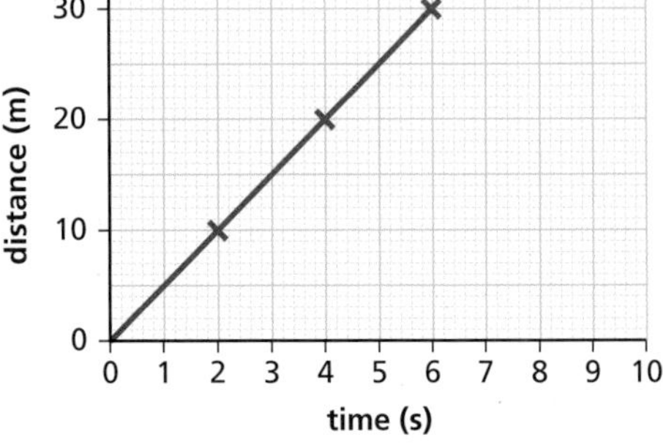

9.7 *Distance–time graph for a faster runner.*

A distance–time graph can tell you a lot about what an object is doing.

Figure 9.8 shows part of a cyclist's journey. In part A, the cyclist moves at a constant speed. In part B he is still moving at a constant speed, but now he is moving faster. In part C he slows down.

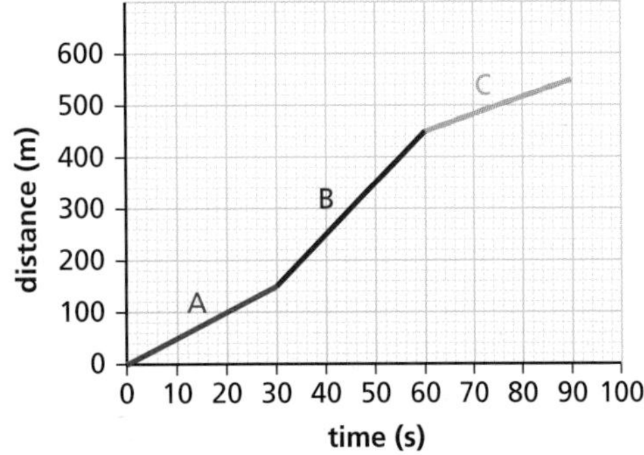

9.8 *Distance–time graph for moving at different average speeds.*

4 Look at figure 9.8.

a) Explain how you know that the cyclist is travelling faster in part B than in part A?

b) Is the cyclist's speed in part C faster or slower than part A? How do you know?

Figure 9.9 is a distance–time graph of an animal that runs at a constant speed for 3 s then stops.

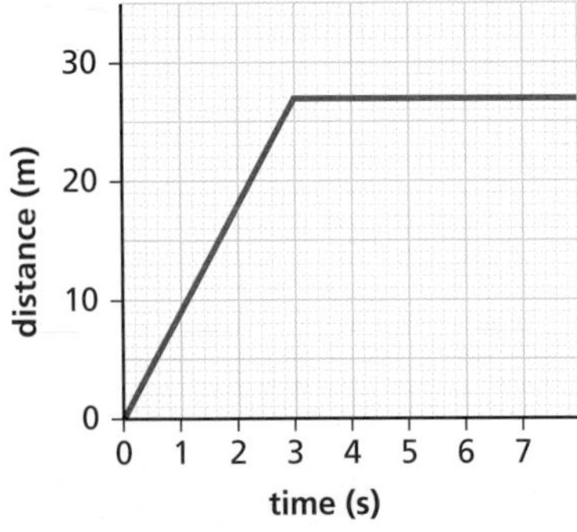

9.9 *Distance–time graph of an animal that runs at a constant speed for 3 s then stops.*

5 How do you know that the animal in figure 9.9 stopped running after 3s?

The graph in figure 9.10 shows a distance–time graph for an animal whose speed is changing all the time.

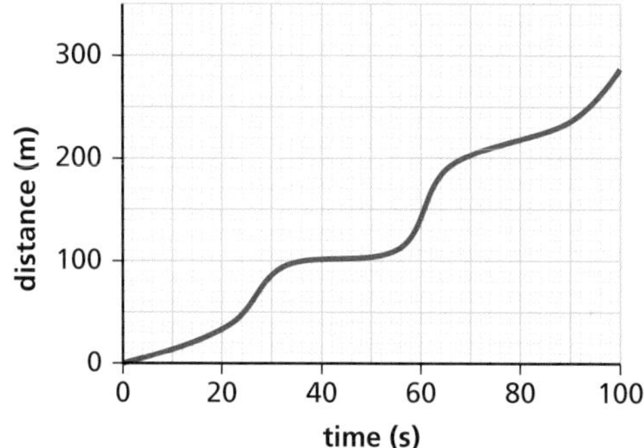

9.10 *Distance–time graph of an animal whose speed is changing all the time.*

6 How can you tell from the graph in figure 9.10 that the animal's speed is changing all the time?

7 Figure 9.11 shows the distance–time graph of a walk. Match the labelled sections (A–E) to the descriptions here.

I turn back towards the start of my walk.
I start to walk at a constant speed.
I feel tired so I slow down.
I start to walk a little quicker.
I start to walk downhill so my speed increases.

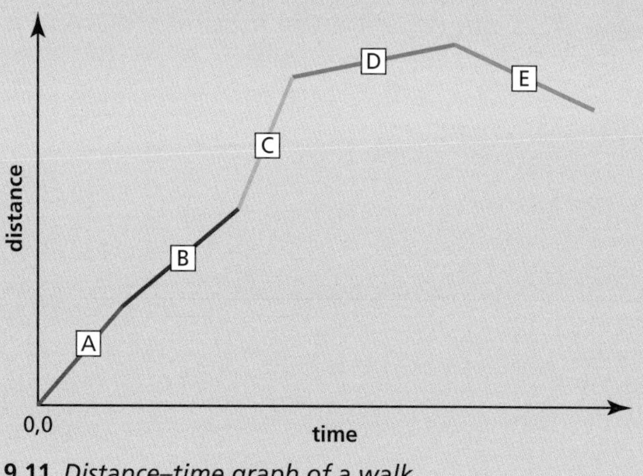

9.11 *Distance–time graph of a walk.*

Plot your own distance–time graph

Activity 9.6: Go on a journey

Work in a small group

A Place four markers on the ground, 10 m apart (see figure 9.12).

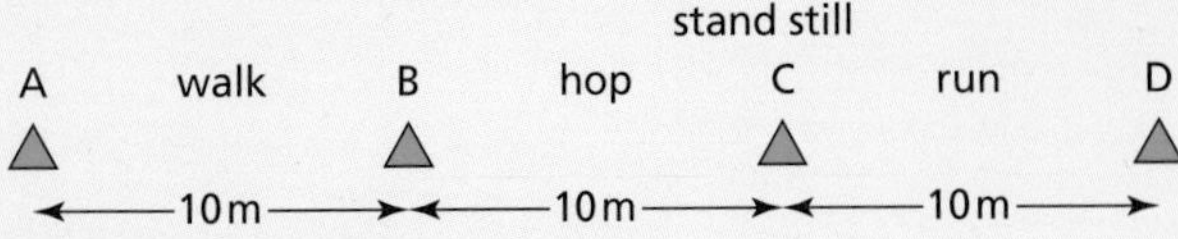

9.12 *Timing a journey.*

B One of you:

- walks from A to B
- then hops from B to C
- then stands still for 5 s
- then runs from C to D.

Time how long each stage of the journey takes. Record the results in a table.

A1 Work out the average speed for each stage of the journey. Draw a distance–time graph of the journey.

Key facts:

✔ The gradient of a distance–time graph tells you the speed of a moving object.

✔ The steeper the gradient of a distance–time graph, the faster an object is moving.

Check your skills progress:

I can draw distance–time graphs for moving objects.

I can describe the motion shown in a distance–time graph.

End of chapter review

Quick questions

1. How far will a train with a speed of 20 m/s travel in:

(a) 10 s? **[1]**

(b) 2 minutes? **[1]**

(c) 10 minutes? **[1]**

2. A dog takes 20 s to run 220 m. What is its average speed? **[1]**

3. How far could a horse with an average speed of 15 m/s run in:

(a) 10 s? **[1]**

(b) 1 minute? **[1]**

4. How long would it take a leopard, running at an average speed of 20 m/s to travel 500 m? **[1]**

5. Speeds can be measured in different units.

(a) What is a speed of 3 m/s equal to in cm/s? **[1]**

(b) What is a speed of 50 cm/s equal to in m/s? **[1]**

(c) What is a speed of 60 km/h equal to in m/s? Give your answer to the nearest whole number. **[1]**

6. What is the average speed of the object shown in the graph below? **[2]**

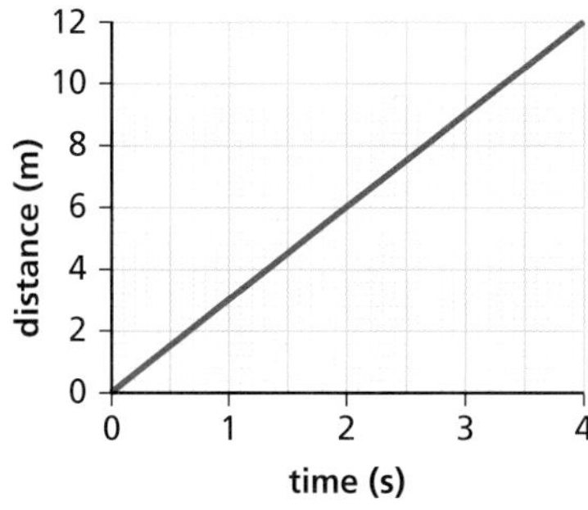

9.13 **[2]**

Connect your understanding

7. Copy the graph in Question 6. Now add a line showing an object moving at an average speed of 2 m/s. **[2]**

8. Why is a light gate a more accurate way of measuring time than a stopwatch? **[2]**

9. The graph below shows the average speed of a racing car at different times as it travels around part of a race circuit. There is a 10 s time interval between each reading. Average readings over each 10 s interval have been plotted.

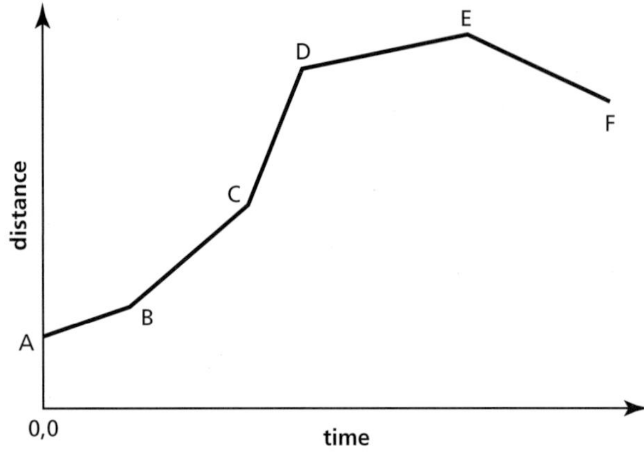

9.14

(a) Does the graph show the start of the race? How do you know? **[1]**

(b) The car slows down when it reaches a corner. Which point on the graph shows where a corner could be? **[1]**

(c) Which point on the graph shows the car at the furthest point from the start of the race? How do you know? **[1]**

(d) Challenge What would the graph look like if the average speed were measured every 1/10th second rather than every 10 seconds? Give a reason for your answer. **[2]**

10. Sketch distance–time graphs for the following:

(a) A car travelling away from its starting point at a constant speed then braking and coming to a stop. **[3]**

(b) A bike moving at a constant speed then suddenly stopping. **[3]**

11. A student wants to find the average speed of a tortoise. She measures how far the tortoise moves every 10 s for 1 minute. Her results are shown in table 9.1.

Time (s)	Distance (m)
0	0
10	4
20	8
30	12
40	15
50	20
60	24

Table 9.1

(a) Draw a distance–time graph of these results. [5]

(b) What is the tortoise's average speed over one minute? [1]

(c) Which one of her readings suggests that the tortoise is not moving at a constant speed? How can you tell this from the graph? [2]

Challenge questions

12. A student is measuring the time it takes a ball to fall from a height of 2 m to the ground. Her results are shown in table 9.2.

Time (s)
0.65
0.67
0.65
0.56
0.66
0.64
0.65

Table 9.2

(a) Which of these results is anomalous? [1]

(b) Calculate the mean time it takes for the ball to fall to the ground. Use all seven results. [1]

(c) Calculate the mean time it takes for the ball to fall to the ground. Use only the results that are not anomalous. [1]

(d) Which of the calculations, (b) or (c), is likely to give the more accurate answer? [1]

(e) Calculate the average speed of the ball as it falls to the ground. Use the more accurate value of mean time taken to reach the ground. [1]

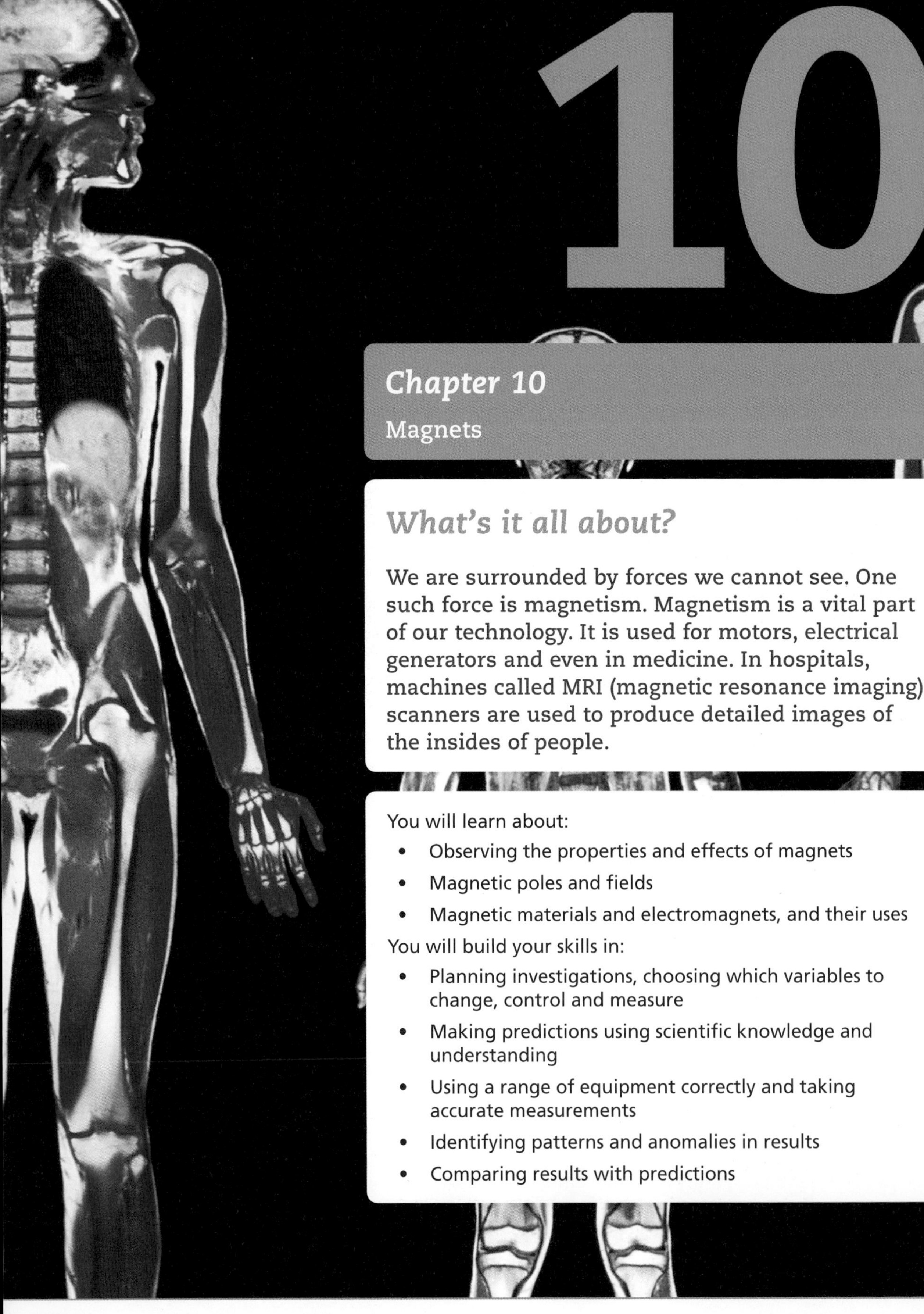

Chapter 10

Magnets

What's it all about?

We are surrounded by forces we cannot see. One such force is magnetism. Magnetism is a vital part of our technology. It is used for motors, electrical generators and even in medicine. In hospitals, machines called MRI (magnetic resonance imaging) scanners are used to produce detailed images of the insides of people.

You will learn about:

- Observing the properties and effects of magnets
- Magnetic poles and fields
- Magnetic materials and electromagnets, and their uses

You will build your skills in:

- Planning investigations, choosing which variables to change, control and measure
- Making predictions using scientific knowledge and understanding
- Using a range of equipment correctly and taking accurate measurements
- Identifying patterns and anomalies in results
- Comparing results with predictions

Magnets and magnetic materials

Learning outcomes

- To describe the properties of magnets
- To investigate the properties of magnets and present results of investigations

Starting point

You should know that...	You should be able to...
Forces change the speed and direction of objects	Outline plans to carry out investigations, considering the variables to control, change or observe
The weight of an object is due to the force of gravity	Make predictions and review them against evidence
The effects of a force can be measured even though the force itself cannot be 'seen'	Make careful observations including measurements
	Make predictions referring to previous scientific knowledge and understanding

Discovery of magnetism

What is magnetism?

If an object changes speed or direction, you already know that a force must be acting. A compass needle turns and eventually stops to point in a particular direction. This means a force must be acting on it.

We say that there is a **magnetic force** acting on the compass. This force is due to the Earth. The Earth itself is a giant magnet!

Key facts about magnetism

Over time, scientists have investigated **magnetism**. They have discovered that:

- Only some materials can be made into **permanent magnets**.
- These materials can be formed into magnets of different shapes, such as bar magnets or horseshoe magnets (see figure 10.1).
- All magnets have **magnetic poles** – a north pole and a south pole. (A north pole or a south pole cannot exist on its own.)

10.1 *A selection of different magnets.*

Key term

magnetism: property of some materials that gives rise to forces between these materials and magnets.

magnetic force: force that occurs when a magnet attracts another object or repels another magnet.

Key terms

magnetic pole: point on a magnet where the force is strongest.

Key terms

permanent magnet: object made from a magnetic material that retains its magnetism for a very long time.

- If the north pole of one magnet is brought close to the south pole of another magnet, the poles attract (see figure 10.2). We say that 'unlike poles **attract**'.
- If the north pole of one magnet is brought close to the north pole of another magnet, the poles repel. We say that 'like poles **repel**'.

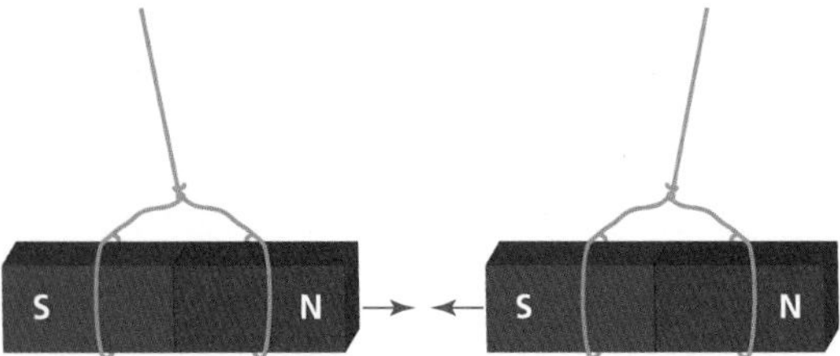

10.2 *If two opposite poles are placed near each other, they attract.*

Key terms

attract: pull closer together.

repel: push further apart.

1 Predict what will happen if the south pole of one magnet is brought close to the south pole of another magnet. Use the key terms to explain your answer.

Activity 10.1: Uses for magnets

Look around your classroom, and think about your possessions and the items around your home. Which items do you think use magnets? Why do you think this?

A1 For each item:

- Describe how you could show whether or not a magnet is involved.
- Suggest how each item uses its magnet.

A2 How strong is the magnet used? Explain your answer. Present your findings in a table.

Magnetic materials

Some materials are affected by magnetic fields. We can show this by bringing a bar magnet close to the material. If it is a magnetic material, it will be attracted to the bar magnet.

Any object made from a magnetic material can be made to behave like a magnet. Slowly stroke a magnetic material with one pole of a bar magnet several times, always starting at the same end (see figure 10.4). When you take the bar magnet away, the object will behave like a weak bar magnet. The effect does not last long.

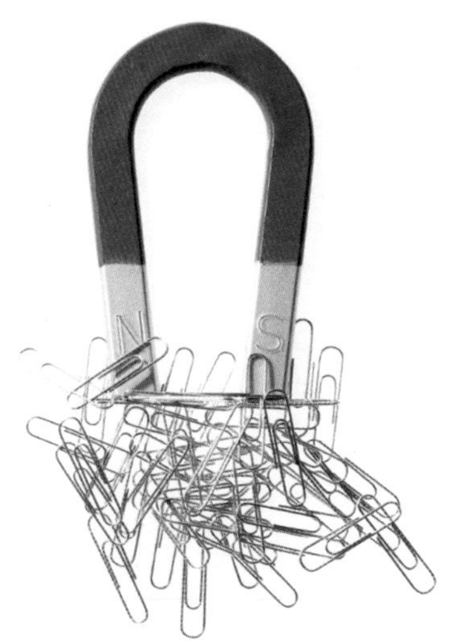

10.3 *These paperclips are made of a magnetic material so they are attracted to the magnet.*

10.4 *You can make this needle magnetic by stroking it with a magnet several times in the same direction.*

Magnetism is a property of the arrangement of the atoms in the structure of a material. We can shape magnetic materials, then magnetise the object using a very strong magnet. The object is then heated to 'fix' the arrangement of the atoms, to make the object a permanent magnet.

If you heat a magnet too much, it will lose its magnetic properties.

Some materials are not affected at all by magnetic forces. See table 10.1.

Activity 10.2: Is this material magnetic?

Test some materials to see whether they are magnetic. You could try iron, nickel, steel, paper, copper, gold, aluminium, card and plastic. Record your results in a table like this one.

Magnetic materials	Non-magnetic materials

Table 10.1 *Magnetic and non-magnetic materials.*

The Earth itself is a giant magnet, which is why it has a north pole and a south pole. It is a giant magnet because it has liquid iron and nickel in its core. The geographical north pole of the Earth is actually a magnetic south pole. We call it the north pole of Earth so that we know that the north poles of compasses point north.

2 You have several different materials. Describe how you would test whether they are magnetic or non-magnetic.

The first compasses

At least 2400 years ago, ancient Greek and Chinese philosophers independently discovered a material that behaved strangely. If a piece of this material was shaped to form a pointer and allowed to spin freely, it would always come to rest pointing in the same direction. This happened no matter where the pointer was placed: indoors or outdoors, in the dark or in the light.

10.5 *An ancient Chinese compass made from lodestone.*

What is a magnetic field?

If you hold a paperclip near a permanent magnet, the material is attracted to the magnet. Even though you cannot see the force, you can feel its effect – you have to exert a force to hold the material in place.

The closer to the magnet you hold the magnetic material, the stronger the force gets. We say that there is a **magnetic field** surrounding the magnet. This field is stronger closer to the magnet, and weaker further away.

Key term

magnetic field: the region around a magnetic material in which a magnetic force acts.

Seeing magnetic field lines

Iron filings are tiny flakes of iron. If you place a piece of white paper on top of a bar magnet and carefully scatter iron filings onto the paper, each filing aligns itself with the magnetic field of the magnet (see figure 10.6).

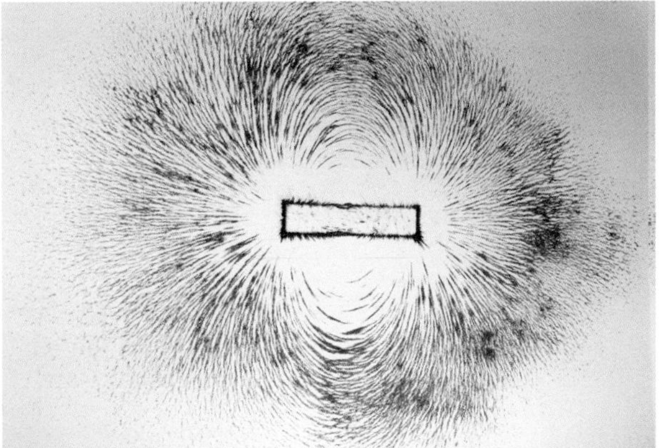

10.6 *Iron filings revealing the magnetic field around a bar magnet.*

Using iron filings in this way gives a good idea of the shape of the magnetic field, but it is not very precise. You can use a plotting compass, pencil and paper to produce a better diagram of the magnetic field (see figure 10.7).

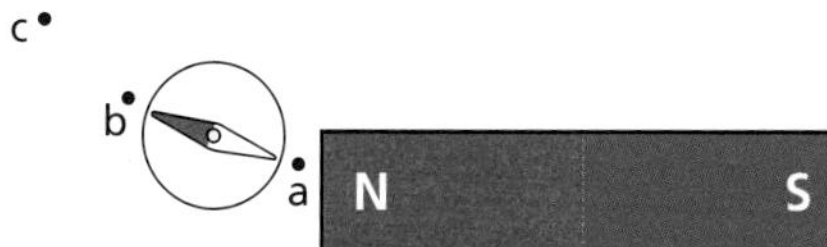

10.7 *How to use a plotting compass to investigate the magnetic field around a bar magnet.*

We can show this field by drawing magnetic field lines (see figure 10.8). The distance between the lines shows the strength of the field: the closer the lines, the stronger the field.

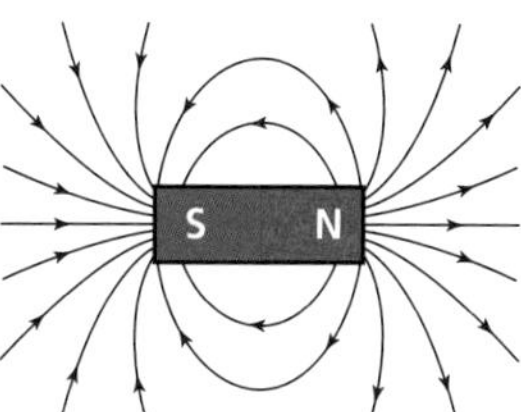

10.8 *The field lines around a bar magnet.*

Drawing magnetic field lines

Remember these rules:

- Each line starts on a north pole and ends on a south pole.
- Each line should be smoothly curved and continuous (a field line must not break in the middle).
- Magnetic field lines cannot cross each other, join together or split apart.
- An arrow on each field line should go from the north pole to the south pole.

Figure 10.8 shows the magnetic field around a simple bar magnet.

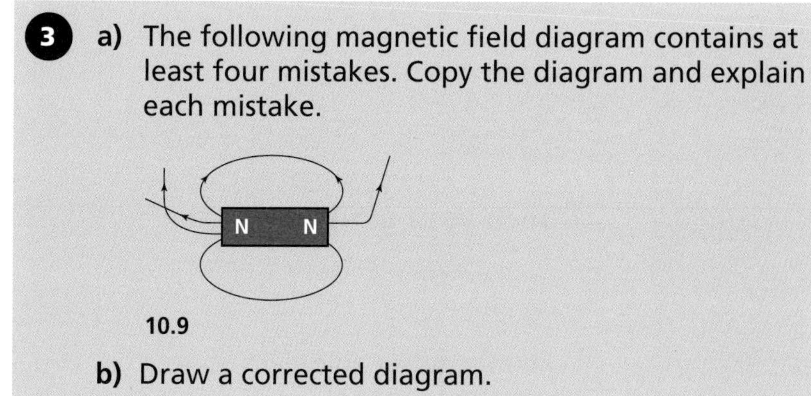

3 **a)** The following magnetic field diagram contains at least four mistakes. Copy the diagram and explain each mistake.

10.9

b) Draw a corrected diagram.

Figure 10.10 shows an experiment to find the field lines around a bar magnet.

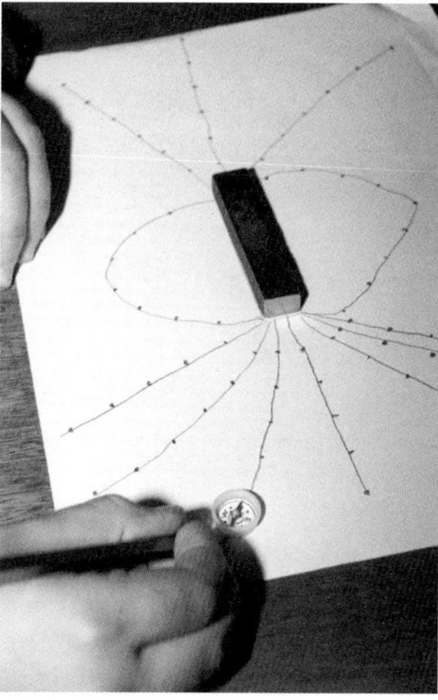

10.10 *An experiment to find the field lines around a bar magnet.*

Method

We can find the field lines around a magnet by using a plotting compass and a piece of paper. First, you place the magnet on the paper and draw around it. Then you place the compass on the paper near the north pole of the magnet. Mark a dot where the needle of the compass points. Then you move the compass so the back of the needle is where the dot you just drew is. The needle may change direction. Mark a dot where the needle points. Repeat this process until you get the field.

Field between two magnets?

Remember:

- like poles repel
- magnetic field lines must not cross or break.

If you bring the north pole of one magnet to face the north pole of a second magnet, the two sets of magnetic field lines bend away from each other (see figure 10.11). The arrangement of field lines here shows that forces will act to push the magnets apart. The closer together the two north poles are, the closer the field lines and the stronger the forces.

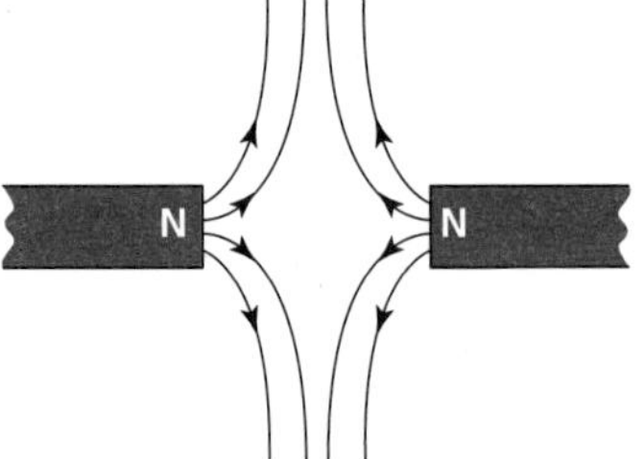

10.11 *The field lines around two magnets that repel because like poles are facing each other.*

4 Describe the forces that act when two opposite poles (one N, one S) of different bar magnets are brought closer together.

5 Draw the magnetic field when two unlike poles are held a short distance apart.

6 Use the field lines to predict what the magnets would do if allowed to move freely.

Key facts:

- ✔ Magnetic materials experience magnetic forces.
- ✔ Every magnet has a north pole (N) and a south pole (S).
- ✔ Magnetic forces can attract or repel.
- ✔ Like poles repel, unlike poles attract.
- ✔ Magnetic field lines show the strength and direction of a magnetic field around a magnetised object.

Check your skills progress:

I can investigate which materials are magnetic.

I can magnetise an object using a permanent magnet.

I can plot and draw magnetic field lines using a plotting compass.

I can predict and explain the behaviour of two magnets when they are placed close together.

Electromagnets

Learning outcomes

- To construct and use an electromagnet
- To identify important variables, choose which variables to change, control and measure

Starting point

You should know that...	You should be able to...
Some metals have magnetic properties	Outline plans to carry out investigations, considering the variables to control, change or observe
Circuits are used to transfer electrical energy from one place to another	Make predictions referring to previous scientific knowledge and understanding
It is possible to measure the effect of a force even though the force cannot be seen	Make predictions and review them against evidence
	Choose appropriate apparatus and use it correctly
	Make careful observations including measurements
	Present results in the form of tables, bar charts or line graphs
	Make conclusions from collected data, including those presented in a table or graph

Electrical current creates a magnetic field

When a current flows through a wire, it creates a magnetic field. You can detect this magnetic field using plotting compasses around the wire.

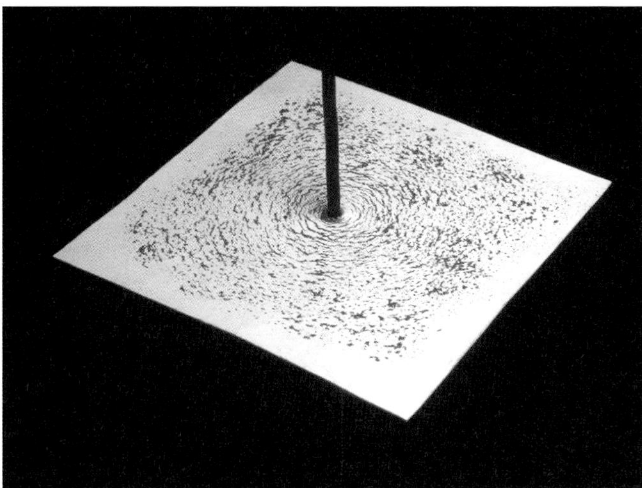

10.12 *The magnetic field around a wire.*

a)  b) 

10.13 *A plotting compass when a) the current is off and b) the current is on.*

Using coils as magnets

If you coil a wire in a cylinder shape and pass a current through it, it creates a field that is similar to a bar magnet. We can coil this wire around a cylindrical piece of metal to increase its strength. We call this piece of metal a **core**. The material a core is made from can affect the strength of the **electromagnet**.

Electromagnets are different to permanent magnets because you can turn an electromagnet on and off. You can also control the strength of an electromagnet. You can do this by changing:

- the number of turns in the coil: more turns create a stronger magnetic field
- the current flowing through the coil of wire: a bigger current creates a stronger magnetic field
- the metal the core is made from: a core made from a magnetic material such as iron creates a stronger magnetic field than a core made from a non-magnetic material, such as plastic.

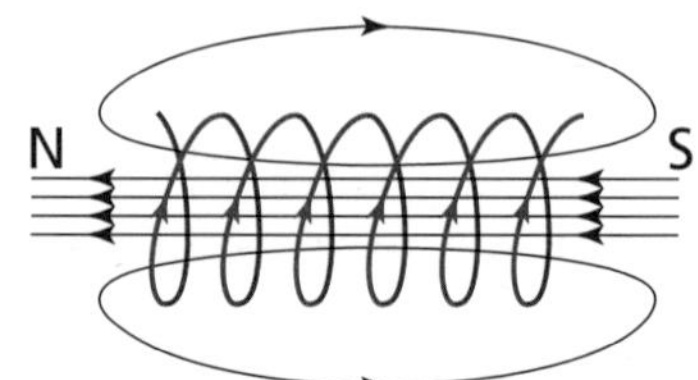

10.14 *The magnetic field lines around a coil of wire.*

10.15 *The wire is wrapped round a solid core.*

Key terms

core: piece of metal (usually iron) that a coil of wire is wound around to increase the strength of the magnetic field.

electromagnet: magnet that can be switched on or off using an electric current.

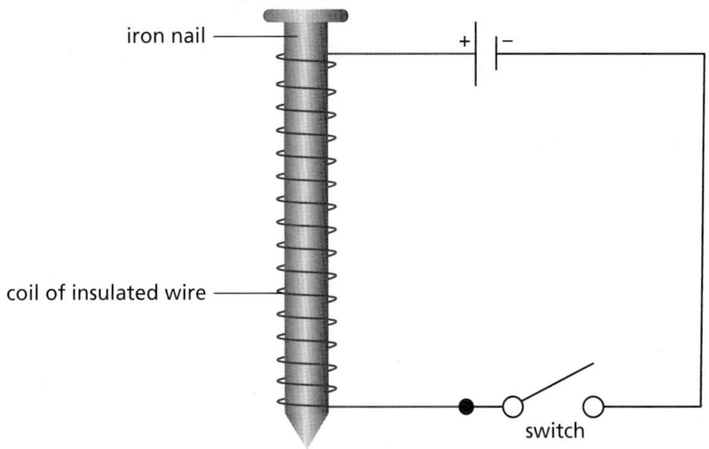

10.16 *The circuit diagram for a simple electromagnet.*

1. Describe how an electromagnet is different from a permanent magnet.
2. Tariq needs to choose a core for his electromagnet. He has iron, copper, plastic and wood. Which material would you recommend? Why?
3. You can change the strength of an electromagnet. Explain why this is useful.

Activity 10.3: What affects the strength of an electromagnet?

The strength of an electromagnet depends on the type of core inside the coil of wire, the number of turns in the coil and the size of the current going through the wire.

Nadia decided to test whether the current affected the strength of an electromagnet. She decided to test how strong it was by counting how many paperclips it could pick up. Nadia decided to use small paperclips and put 20 coils around an iron core.

Nadia predicted that as she increased the current, the electromagnet would pick up more paperclips.

Here are Nadia's results:

Current (amps)	Number of paperclips the electromagnet picks up.
1	10
2	21
3	33
4	45
5	56

Table 10.2 *Nadia's results.*

A1 Use the key terms to identify the **independent variable** in this investigation.

A2 Use the key terms to identify the **dependent variable** in this investigation.

A3 Use the key terms to identify three **control variables** in this investigation.

A4 Draw a graph of your results to show what the trend is like.

A5 Do the results agree with Nadia's prediction? Explain your answer.

4 Mira did an investigation on how the number of coils around an electromagnet affected the number of paperclips it picked up. She collected some results. Plot a graph of her results.

Number of coils	Number of paperclips I picked up
0	0
5	1
10	3
15	3
20	8
25	12

Table 10.3 *Mira's results.*

Key terms

control variable: the variable in an investigation that you keep the same.

dependent variable: the variable in an investigation that you measure.

independent variable: the variable in an investigation that you change.

a) One of the results is anomalous. Which one?

b) Plot a line of best fit on your graph, but don't include the anomalous result.

c) What conclusion can Mira draw from the results?

5 Lee changed both the number of coils and the current when he did his investigation. When he increased them both, he picked up more paperclips. Why can't he conclude that the number of coils affects the strength of the electromagnet?

Uses of electromagnets

Industrial lifting magnets

Large electromagnets are used in scrapyards to pick up and move large objects made from magnetic material such as steel and move them to other places. You can lift something up when the current is on, and drop it by turning the current off.

10.17 *An industrial lifting magnet.*

Electric bells

Electric bells use electromagnets. When you press the switch, you complete a circuit, and so turn on the electromagnet. This attracts the striker to the bell which makes the noise. However, when the striker strikes the bell, it breaks the circuit which turns off the electromagnet. The striker goes back to its original place, but this turns on the electromagnet again and attracts the striker again. This continues until you stop pressing the switch.

Relays

A relay is a device that turns on a circuit with a high current using a circuit with a low current. This is important as it stops you being too close to high current and so makes you safer. When you push a switch on a relay, it turns on an electromagnet, which attracts a piece of iron connected to another circuit. This circuit is now complete so that begins to work. Relays are used in cars so that you can start a motor safely.

MRI scanners

Magnetic resonance imaging (MRI) is a very useful tool in medicine. With MRI, the patient is put into a strong magnetic field created by an electromagnet. In this field, the atoms in the patient's body act like tiny magnets and align with the field. This can be used to produce an image of the inside of the body which can be used to detect many conditions such as tissue damage, diseases affecting the brain, some cancers, heart disease and diseases in bone joints. Since they only use a magnetic field, MRI scanners are much safer than other techniques such as X-ray scanning and you can see tissue as well as bone.

Maglev trains

Maglev trains use electromagnets to float above the rails. They are used in Japan, China and South Korea. Maglev is short for magnetic levitation. These trains have no moving parts and there is no friction between the train and the track. This means that they are faster and quieter than conventional trains. They also cheaper to maintain after they are built.

10.18 *A maglev train.*

There are electromagnets on the train and the track. Some maglev trains have electromagnets that repel each other to make them levitate. Others have electromagnets that are attracted to electromagnets underneath the track and this makes them levitate.

6. Why can't you make an electric bell using a permanent magnet?

7. Maglev trains have very small maintenance costs. Give a reason why you think people haven't built more of them.

Key facts:

- ✔ When a current flows through a wire, it creates a magnetic field.
- ✔ When a wire is coiled into a cylinder shape and a current is passed through it, it has a magnetic field like a bar magnet. This is called an electromagnet.
- ✔ The strength of an electromagnet is affected by the number of coils it has, the current and the type of core used to wrap the wires around.
- ✔ Electromagnets have many uses because they can be turned on or off and their strength can be changed.

Check your skills progress:

I can construct and use an electromagnet.

I can use a range of equipment to discuss what affects the strength of an electromagnet.

I can present my results in a table and graph.

I can use my scientific understanding to discuss and explain what affects the strength of an electromagnet.

End of chapter review

Quick questions

1. Give the definitions for the following terms:

   (a) permanent magnet [1]

   (b) electromagnet [1]

   (c) magnetic material [1]

   (d) magnetic field. [1]

2. Mariam has some pieces of metal. She does not know if each piece of metal is a permanent magnet, a magnetic material or non-magnetic material. For each piece of metal, say if it is a permanent magnet, a magnetic material or not magnetic. Give a reason for your answer.

   (a) One metal is attracted to both poles of a magnet. [2]

   (b) One substance does not move at all when a magnet is placed near it. [2]

   (c) One substance is attracted to one end of a magnet and repelled by the other end. [2]

3. Which of the following mixtures can be separated by a magnet and why?

   (a) Iron and nickel [2]

   (b) Steel and plastic [2]

   (c) Wood and copper [2]

4. Give instructions for turning a steel needle into a permanent magnet. [2]

5. Look at the diagrams below. Copy the diagrams and draw an arrow to show which direction the blue magnet will move in.

   (a) 

[1]

   (b) 

[1]

   10.19

6. Copy the diagram below. Add arrows to the plotting compasses to show which direction they point.

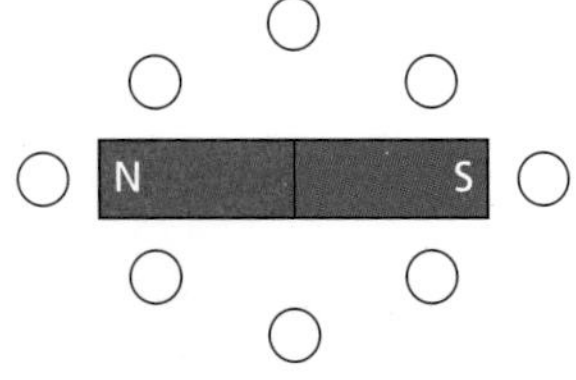

10.20 [8]

7. Describe a method, other than using plotting compasses, to see a magnetic field. **[2]**

8. Copy the diagrams below. Draw fields around the magnets.

(a) **[2]**

10.21

(b) **[2]**

10.22

9. Which pole of a magnet points to the north pole of the Earth? Explain your answer. **[2]**

Connect your understanding

10. Omar investigated how the number of coils around an iron core affects the strength of an electromagnet. He measured the strength by measuring how many paperclips it picked up. He obtained the results given in table 10.4.

Number of coils	Number of paperclips picked up
0	0
10	3
20	12
30	19
40	25
50	30
60	34

Table 10.4 *Omar's results.*

Plot a graph of his results. **[3]**

11. Here are magnetic fields around two pairs of magnets.

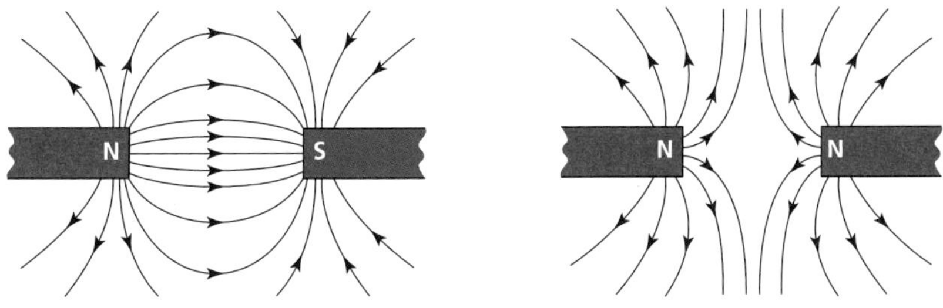

10.23

(a) Which two magnets are attracted? **[1]**

(b) Which two magnets are repelling? **[1]**

(c) Draw the field around two magnets with south poles facing each other. **[1]**

Challenge questions

12. Sarah did an investigation into how the type of core in an electromagnet affects its strength. She measured how many paperclips each picked up. She obtained the results shown in table 10.5.

Core type	Number of paperclips picked up
Air	1
Copper	4
Iron	7

Table 10.5 *Sarah's results.*

(a) What was the independent variable in this investigation? **[1]**

(b) What was the dependent variable in this investigation? **[1]**

(c) Name one control variable in this investigation. **[1]**

(d) Steel is a mixture of iron and carbon. Predict how many paperclips a steel core could pick up. Explain your answer. **[2]**

13. Use the diagram to explain how an electric bell works. **[5]**

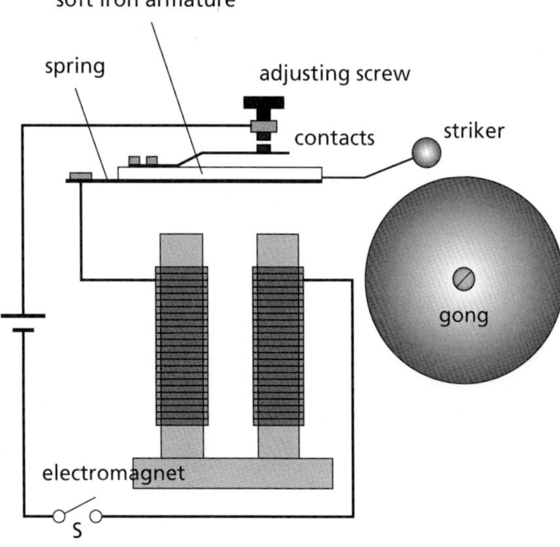

10.24

End of stage review

1. (a) Zikri has three torches:

- a torch that shines pure green light
- a torch that shines pure blue light
- a torch that shines pure red light.

Copy and complete these sentences. Choose from the following words:

black blue cyan green magenta red yellow white

(i) If light from the red torch overlaps light from the green torch, ______________ light is produced. **[1]**

(ii) Zikri knows that magenta light is made from red and blue light. If he shines magenta light through a red filter ______________ light passes though the filter. **[1]**

(b) Zikri shines the red torch onto a blue towel. Explain why the towel looks black. **[2]**

(c) Marta looks at a coin in a beaker of water. The real depth of the coin is 20 cm. Write the letters of the *two* true statements? **[2]**

A The apparent depth is not the same as the real depth because of reflection.

B The apparent depth is not the same as the real depth because of refraction.

C The apparent depth of the coin is less than 20 cm.

D The apparent depth of the coin is more than 20 cm.

2. (a) Mia stretches an elastic band using a hook and a ruler. When she pulls the elastic band with her finger it vibrates.

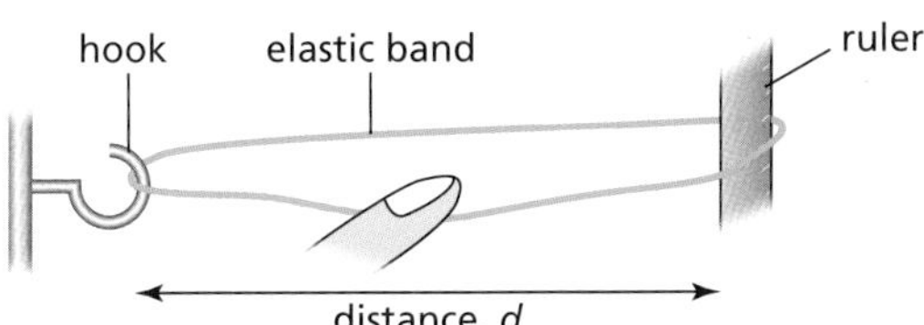

Copy and complete the sentences. Choose *one* of these words to complete each sentence. You can use a word more than once.

higher lower

(i) Mia increases the distance *d* by 2 cm. The pitch of the sound becomes ____________. **[1]**

(ii) A lower pitch sound has _________ frequency. **[1]**

(b) Mia uses an oscilloscope to record three sounds.

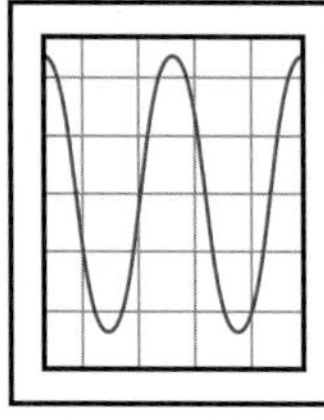

Sound A

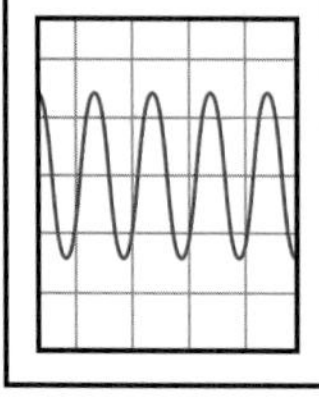

Sound B

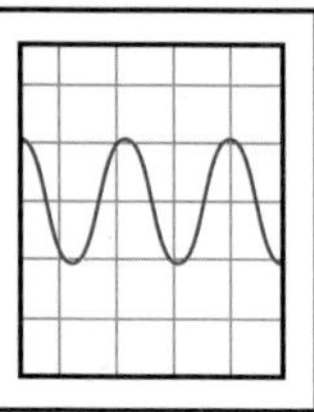

Sound C

(i) Write the name of the sound that is the loudest. **[1]**

(ii) Write the name of the sound that has the highest frequency. **[1]**

(c) Ali is in a thunderstorm.

(i) Explain why Ali see's the flash of lightning before he hears the thunder. **[1]**

(ii) Ali measures the time from seeing the lightning to hearing the thunder

	How long until Ali hears the thunder
1st flash of lightning	15 seconds
2nd flash of lightning	10 seconds
3rd flash of lightning	2 seconds
4th flash of lightning	7 seconds

Write down which flash of lightning is closest to him. **[1]**

3. Sarah walks to school. On the way she stops to talk to a friend. The graph shows her journey.

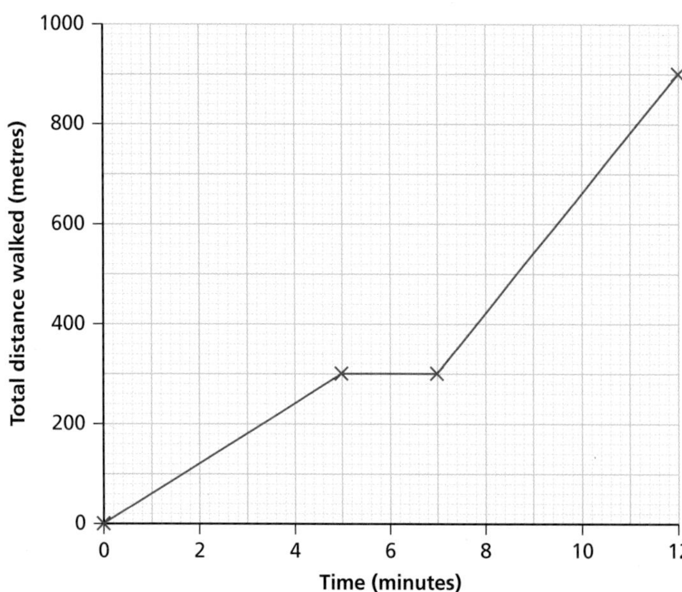

(a) Copy the table below. Use the graph to complete the table.

	Time	Distance walked
First part of journey	________ minutes	300 metres
Talks to a friend	________ minutes	0 metres
Final part of journey	5 minutes	________ metres

[3]

(b) Calculate Sarah's average speed in metres/second from home to school. Show how you worked out your answer. **[2]**

(c) Explain how you know from the gradient of the graph that Sarah walked faster on the final part of her journey. **[1]**

4. (a) Copy and complete these sentences. Choose one of these words to complete each sentence.

attract repel

(i) These two magnets ___________ each other [1]

(ii) These two magnets ___________ each other. [1] **[1]**

(b) Write the letter of the correct drawing of the field lines around a bar magnet.

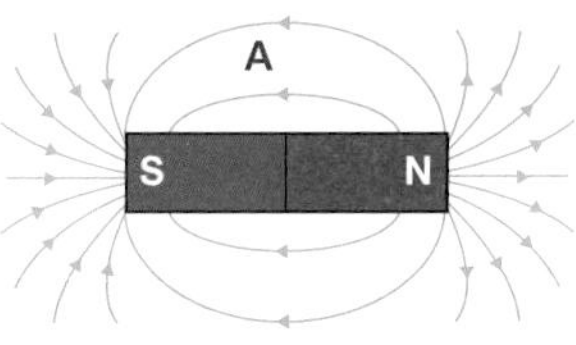

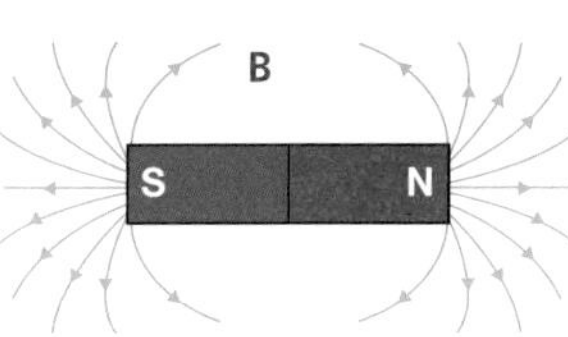

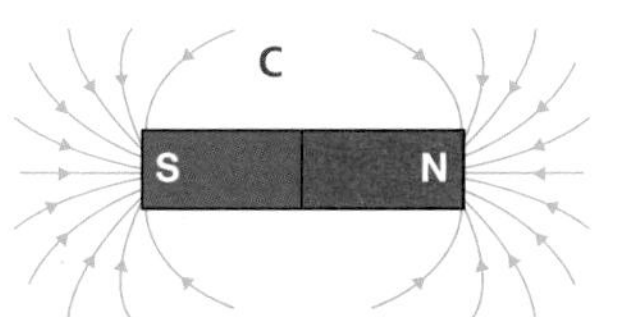

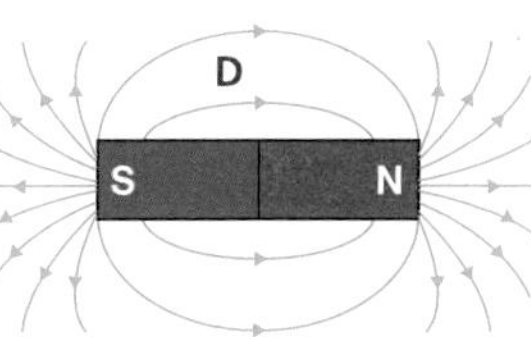

[1]

(c) Explain how Abhi could use an electromagnet to separate these aluminium and steel cans into separate containers. **[2]**

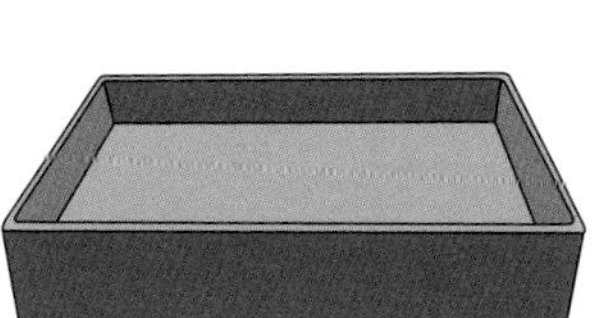

[total 22 marks]

Periodic Table

Key

relative atomic mass **atomic symbol** name atomic (proton) number

1	2											3	4	5	6	7	0
							1 **H** hydrogen 1										4 **He** helium 2
7 **Li** lithium 3	9 **Be** beryllium 4											11 **B** boron 5	12 **C** carbon 6	14 **N** nitrogen 7	16 **O** oxygen 8	19 **F** fluorine 9	20 **Ne** neon 10
23 **Na** sodium 11	24 **Mg** magnesium 12											27 **Al** aluminium 13	28 **Si** silicon 14	31 **P** phosphorus 15	32 **S** sulfur 16	35.5 **Cl** chlorine 17	40 **Ar** argon 18
39 **K** potassium 19	40 **Ca** calcium 20	45 **Sc** scandium 21	48 **Ti** titanium 22	51 **V** vanadium 23	52 **Cr** chromium 24	55 **Mn** manganese 25	56 **Fe** iron 26	59 **Co** cobalt 27	59 **Ni** nickel 28	63.5 **Cu** copper 29	65 **Zn** zinc 30	70 **Ga** gallium 31	73 **Ge** germanium 32	75 **As** arsenic 33	79 **Se** selenium 34	80 **Br** bromine 35	84 **Kr** krypton 36
85 **Rb** rubidium 37	88 **Sr** strontium 38	89 **Y** yttrium 39	91 **Zr** zirconium 40	93 **Nb** niobium 41	96 **Mo** molybdenum 42	[98] **Tc** technetium 43	101 **Ru** ruthenium 44	103 **Rh** rhodium 45	106 **Pd** palladium 46	108 **Ag** silver 47	112 **Cd** cadmium 48	115 **In** indium 49	119 **Sn** tin 50	122 **Sb** antimony 51	128 **Te** tellurium 52	127 **I** iodine 53	131 **Xe** xenon 54
133 **Cs** caesium 55	137 **Ba** barium 56	139 **La*** lanthanum 57	178 **Hf** hafnium 72	181 **Ta** tantalum 73	184 **W** tungsten 74	186 **Re** rhenium 75	190 **Os** osmium 76	192 **Ir** iridium 77	195 **Pt** platinum 78	197 **Au** gold 79	201 **Hg** mercury 80	204 **Tl** thallium 81	207 **Pb** lead 82	209 **Bi** bismuth 83	**Po** polonium 84	**At** astatine 85	**Rn** radon 86
Fr francium 87	**Ra** radium 88	**Ac**** actinium 89	**Rf** rutherfordium 104	**Db** dubnium 105	**Sg** seaborgium 106	**Bh** bohrium 107	**Hs** hassium 108	**Mt** meitnerium 109	**Ds** darmstadtium 110	**Rg** roentgenium 111	Elements with atomic numbers 112–116 have been reported but not fully authenticated						

La lathanoids
Ac actinoids

Elements 1 to 92 are naturally occurring elements on Earth. Elements 93 and above are man-made

Glossary

Biology

absorb: to take in or soak up.

addictive: substance that makes people feel that they must have it.

adolescence: the life stage in humans that usually happens between the ages of 11 and 18. During this stage, people go through many emotional and physical changes.

aerobic respiration: respiration that requires oxygen to release energy from glucose.

alimentary canal: tube that runs from your mouth to your anus.

alveolus: tiny, pocket-shaped structure in lungs where gaseous exchange happens. The plural is alveoli.

amniotic fluid: liquid that surrounds the foetus in the uterus.

anus: last organ in the alimentary canal. Faeces leave your body here.

aorta: large artery that leaves the left ventricle of your heart.

artery: thick-walled blood vessel that carries blood away from the heart.

atria: chambers at the top of your heart. You have a left atrium and a right atrium.

balanced diet: eating many different foods to get the correct amounts of nutrients.

bile: liquid that helps fat-digesting enzymes to work.

biomass: mass of all the compounds in an organism, which it has made.

blood: liquid organ that carries substances around the body.

blood vessels: tube-shaped organs that carry blood around the body.

breathing: movements of muscles in your respiratory system that cause air to move in and out of your lungs.

breathing rate: the number of times you inhale and exhale in one minute.

bronchioles: small tubes leading from the bronchus in a lung.

bronchus: large tube leading from the trachea into a lung. Plural is bronchi.

cancer: when cells in a tissue start to make many copies of themselves very quickly.

capillary: tiny blood vessel that carries blood from arteries to veins.

carbohydrate: compound made from carbon, hydrogen and oxygen.

catalyst: substance that speeds up a chemical reaction.

cervix: the neck of the uterus.

chamber: space inside the heart that fills with blood and empties again.

chemical digestion: digestion that is done by chemical substances.

chemical reaction: change in which new substances are produced.

chest: area inside the body between the ribcage, neck, backbone and diaphragm.

chlorophyll: green substance that absorbs light, to get energy for photosynthesis.

chloroplast: green part of a plant cell that contains chlorophyll.

cilia: waving strands that stick out of some cells.

ciliated epithelial cell: specialised cell with waving cilia to sweep mucus along.

circulation: movement of blood around the body.

circulatory system: group of organs that gets blood around the body.

clot: thick mass of blood cells, stuck together.

compound: substance made from elements.

constipation: when your intestines become blocked.

contract (muscle): when muscle tissue gets shorter and fatter, it contracts.

Glossary

control variable: variable that you keep the same during an investigation.

correlation: relationship (link) between variables where one increases or decreases as the other increases.

cuticle: waterproof covering on leaves.

deficiency disease: another term for nutritional deficiency.

diaphragm: organ that helps breathing.

diet: what you normally eat or drink.

diffusion: the spreading out of particles from where there are many (high concentration) to where there are fewer (lower concentration).

digestive juice: liquid that contains enzymes to digest food.

double circulatory system: circulatory system in which the heart pumps blood around two circuits. In humans, one circuit supplies the lungs, and the other supplies the rest of the body.

drug: substance that affects the way your body works.

element: substance that contains only one type of atom; it cannot be split into anything simpler.

embryo: small ball of cells that develops from a fertilised egg cell. It becomes attached to the uterus lining and develops into a foetus.

enzyme: substance that digests food.

epidermis cell: cell that forms an outer covering of a leaf, to protect the leaf.

excrete: getting rid of wastes made inside an organism.

exhale: breathing out.

faeces: solid waste material produced by humans and other animals.

fats: nutrients needed by your body to store energy.

fertilisation: when an egg cell nucleus and a sperm cell nucleus fuse (join) and form a fertilised egg cell.

fibre: food substance that cannot be digested but which helps to keep your intestines healthy.

foetus: baby developing in a woman's uterus, from about 10 weeks of development, when it starts to resemble a baby.

gall bladder: organ next to your liver that stores bile.

gamete: sex cell (egg cell in a female and sperm cell in a male).

gaseous exchange: when two or more gases move from place to place in opposite directions.

gestation period: the amount of time for a baby to fully develop in the mother's uterus. This time varies for different animals – for example, 31 days for rabbits and 22 months for elephants.

glucose: sugar made by digesting carbohydrates (in animals) and by photosynthesis (in plants).

growth hormone: chemical made in the brain that causes growth in the body.

guard cell: cell that helps form a stoma in a leaf, to allow gases in and out.

gullet: another word for oesophagus.

gut: another word for alimentary canal.

haemoglobin: substance that collects oxygen.

hazard: harm that something may cause.

heart: organ that pumps blood through blood vessels.

heart attack: when heart muscle cells start to die and the heart does not pump properly.

heartbeat: squeezing of the muscles in the heart wall to push blood into blood vessels.

heart rate: the number of heartbeats in one minute.

high blood pressure: when blood puts too much pressure on blood vessels.

hormone: chemical released into the bloodstream, which has an effect on certain parts of your body.

illegal drug: drug that individual people are not allowed to buy or use. Different countries have different laws about drugs.

implantation: the developing embryo attaches to the uterus lining.

infant: the life stage which lasts for the first year after birth.

inhale: breathing in.

internal fertilisation: when the nuclei of the egg and sperm join together inside the female animal.

iodine solution: liquid that turns from orange to blue-black when added to starch.

joule: unit used to measure energy.

kwashiorkor: deficiency disease caused by a lack of protein.

large intestine: organ of the alimentary canal. It removes water from undigested food to make faeces.

line of best fit: straight or curved line drawn through the middle of a set of points to show the pattern of data points.

lipids: another word for fats.

liver: organ that makes and destroys substances. It makes bile.

lungs: organs that get oxygen into the blood and remove carbon dioxide.

malnutrition: when a diet contains too much or too little of something, and causes health problems.

mechanical digestion: digestion that is done by physical actions, such as chewing.

menopause: the time in a woman's life when the menstrual cycle stops, normally between ages 45 and 55.

menstrual cycle: cycle of changes that happens in females after puberty. During each cycle an egg cell is released from an ovary and (if it is not fertilised) menstruation happens. Each cycle lasts between about 24 and 35 days.

menstruation: the time in the menstrual cycle when the uterus lining breaks down and is lost. It is called a period.

microscopic: something so small you can only see it using a microscope.

mineral salt: type of substance in the soil that plants need small amounts of. Often just called a 'mineral'.

minerals: nutrients that living organisms need in small amounts for health, growth and repair. Also called mineral salts.

mitochondria: part of a cell where aerobic respiration happens to release energy. The singular form is 'mitochondrion'.

model: simple way of showing or explaining a complicated object or idea.

molecule: group of two or more atoms joined together. Oxygen, carbon dioxide and water all exist as molecules.

mouth: first organ in the alimentary canal.

mucus: sticky liquid that traps particles.

nicotine: addictive drug in tobacco smoke.

nutrient: substance you need in your diet for energy or as a raw material.

nutritional deficiency: problem caused by a lack of a nutrient in the diet. Also called a deficiency disease.

obesity: being so overweight that your health is in danger.

oesophagus: organ of the alimentary canal. Its muscle walls push food from your mouth into your stomach.

oestrogen: hormone that triggers many of the physical changes in girls during puberty.

ovary: female reproductive organ where eggs are made, stored and matured. Females have two ovaries.

oviduct: the tube which connects the ovary to the uterus. Females have two oviducts, one from each ovary.

ovulation: the release of an egg cell from one of the ovaries.

palisade cell: cell that contains many chloroplasts for photosynthesis.

pancreas: organ that makes enzymes to digest fats, proteins and carbohydrates.

penis: male reproductive organ used to transfer sperm to female cervix during intercourse.

period: stage in the menstrual cycle when the lining of the uterus is lost from the body.

peristalsis: contraction and relaxation of muscles in the alimentary canal that pushes food along.

pharmaceutical drug: drug used in healthcare to help the body fight a disease, or to relieve pain.

phloem cell: plant cell that is adapted to form living tubes to transport sugars and other substances.

photosynthesis: chemical reaction that plants use to make their own food.

placenta: organ which forms in the uterus, linking the developing foetus to the uterus wall (and therefore the mother).

plaque: lump of fatty material that builds up inside an artery.

plant vein: tube containing smaller tubes that carry substances around a plant.

plasma: liquid part of the blood.

platelet: cell fragment that helps your blood to clot.

pregnant: a woman becomes pregnant if a fertilised egg implants in her uterus.

product: substance made during a chemical reaction.

prostate gland: gland that surrounds the bottom of a male's bladder. It produces some of the liquid that makes up semen.

proteins: nutrients you need for growth and repair.

puberty: the physical changes that happen to the body during adolescence.

pulse: wave of stretching along the wall of an artery each time the heart beats.

raw material: another term for reactant.

reactant: substance that changes in a chemical reaction to form products.

rectum: organ of the alimentary canal. It stores faeces.

red blood cell: cell that contains haemoglobin so it can carry oxygen.

relax (muscle): when muscle tissue stops contracting, it relaxes.

respiration: chemical process that happens in all parts of an organism to release energy.

respiratory system: group of organs that get oxygen into the blood and remove carbon dioxide.

rib: bone that helps to protect your heart and lungs.

ribcage: all your ribs.

rickets: deficiency disease caused by a lack of calcium or vitamin D.

risk: chance of a hazard causing harm.

root hair cell: plant cell found in roots that is adapted for taking in water quickly.

saliva: digestive juice made by salivary glands in your mouth.

salivary gland: organ inside your mouth that makes saliva.

scatter graph: graph of two variables, both measured in numbers.

scurvy: deficiency disease caused by a lack of vitamin C.

secondary sexual characteristics: the physical changes that happen to a person's body during puberty.

semen: the liquid containing sperm.

sexual reproduction: the type of reproduction involving male and female gametes coming together.

small intestine: organ of the alimentary canal. It makes enzymes and lets digested food pass into your blood.

sperm duct: the tube that carries sperm from the testes to the urethra.

spongy cell: irregularly shaped cell that helps form air spaces in a leaf.

starch: large carbohydrate, which plants use to store energy.

stillborn: the term used to describe a baby that is dead when it is born.

stoma: hole in a leaf, formed between two guard cells. The plural is stomata.

stomach: organ of the alimentary canal. It makes enzymes to digest proteins and churns food into a smooth soup.

stroke: when brain cells die due to a lack of blood (which is caused by a blocked blood vessel in the brain).

sugar: type of small carbohydrate.

surface area: the area of a surface, measured in squared units such as square centimetres (cm^2).

symptom: effect of a disease on the body.

tar: sticky black liquid found in cigarette smoke.

testis: (plural testes) male organ where sperm are made. Males have two testes.

testosterone: hormone that triggers many of the physical changes in boys during puberty.

tissue: group of cells of the same type.

trachea: tube-shaped organ that allows air to flow in and out of your lungs.

transpiration: the flow of water into a plant's root, up its stem and out of its leaves.

trend: pattern seen in data.

tumour: a lump of cancer cells.

type 2 diabetes: disease that may damage organs.

umbilical cord: flexible tube containing blood vessels from the foetus – it connects the foetus to the placenta.

urea: waste product made by the liver and excreted by the kidneys.

uterus: female reproductive organ where the baby grows when a woman is pregnant.

vagina: the tube joining the uterus to the outside of the female body.

valve: flaps of tissue that only allow blood to flow in one direction.

variable: something that may change in an experiment.

vein: thin-walled blood vessel that carries blood towards the heart.

ventricle: chamber at the bottom of your heart. You have a left ventricle and a right ventricle.

vitamins: nutrients you need in small amounts for health, growth and repair.

white blood cell: cell that helps destroy microorganisms.

wilting: when a plant becomes floppy due to lack of water.

word equation: model showing what happens in a chemical reaction, with reactants on the left of an arrow and products on the right.

xylem cell: plant cell that is adapted to form hollow tubes to transport water.

zygote: fertilised egg cell.

Glossary

Chemistry

alloy: mixture of metal with other elements.

atom: the smallest particle of a substance (element) that can exist and still be the same substance.

carbonate: compound that reacts with an acid to give carbon dioxide, a salt and water. For example, calcium carbonate ($CaCO_3$).

chemical reaction: a change in which new substances are produced.

chemical symbol: short way of representing an element's name.

chloride: salt that is formed when hydrochloric acid reacts with another element; for example, sodium chloride ($NaCl$).

combustion: chemical reaction between a substance and oxygen, which transfers energy as heat and light.

compound: contains atoms or more than one element strongly held together. Compounds have different properties to the elements they contain.

concentration: measure of how many particles of a certain type there are in a volume of liquid or gas.

correlation: relationship (link) between variables where one increases or decreases as the other increases.

corrosion: the damage of metals through chemical reactions with substances in the air or water.

dense: has a high mass in a small volume.

diffusion: the spreading out of particles from where there are many (high concentration) to where there are fewer (lower concentration).

distillation: separation method used to separate a liquid from a mixture.

ductile: able to be stretched into wires.

element: substance that contains only one type of atom; it cannot be split into anything simpler.

evaporation: method used to separate a soluble solid from a liquid.

filtration: method used to separate an insoluable solid from a liquid.

formula: shows the chemical symbols of elements in a compound, and how many of each type of atom there are.

gas pressure: the effect of the forces caused by collisions from gas particles on the walls of a container.

hazard: harm that something may cause.

hydroxide: compound that contains one atom each of oxygen and hydrogen bonded together; for example, potassium hydroxide (KOH).

insoluble: substance that does not dissolve.

magnetic: material that is attracted by a magnet.

mixture: two or more elements or compounds mixed together. They can easily be separated.

molecule: group of two or more atoms joined together. Oxygen, carbon dioxide and water all exist as molecules.

oxidation: chemical reaction with oxygen to form a compound that contains oxygen.

oxide: compound that is formed when oxygen reacts with another element; for example, magnesium oxide (MgO).

Periodic Table: list of all the elements.

product: substance made during a chemical reaction.

pure: substance that contains only one element or compound.

rate: measurement of how quickly something happens.

reactant: substance that changes in a chemical reaction to form products.

reactivity: how likely it is that a substance will undergo a chemical reaction.

risk: chance of a hazard causing harm.

rusting: chemical reaction of iron with oxygen and water. It is the corrosion of iron.

salt: compound formed when an acid reacts with a base or a metal.

soluble: substance that dissolves to form a solution.

solution: mixture of a soluble substance and a liquid.

sulfate: salt that is formed when sulfuric acid reacts with another element; for example, copper sulfate ($CuSO_4$).

theory: idea or set of ideas that explains an observation.

word equation: model showing what happens in a chemical reaction, with reactants on the left of an arrow and products on the right.

Glossary

Physics

absorption: the way in which an object takes in the energy reaching its surface.

accurate: accurate measurement is where the measurement is very close to the real value.

amplitude: the maximum height of the wave, from the centre to the top or bottom.

angle of incidence: this is the angle between the incident ray and the normal.

angle of reflection: this is the angle between the reflected ray and the normal.

anomalous result: result that does not follow the same pattern as other measurements.

apparent depth: how deep something appears to be.

attract: pull closer together.

average: the mean average of a set of numbers is found by:

$$\frac{\text{total of all the numbers added together}}{\text{how many different numbers there are.}}$$

compression: region where particles in a longitudinal wave are closer together.

control variable: the variable that you keep the same during an investigation.

core: piece of metal (usually iron) that a coil of wire is wound around to increase the strength of the magnetic field.

correlation: relationship (link) between variables where one increases or decreases as the other increases.

dependent variable: the variable in an investigation that you measure.

diffraction grating: transparent piece of glass or plastic which has many lines drawn onto it. Light can pass through the spaces between the lines.

dispersion: the splitting of white light into a spectrum of colours.

electromagnet: magnet that can be switched on or off using an electric current.

filter: colour filter will only allow light of its own colour to pass through it.

frequency: the number of waves per second.

gradient: the gradient of a graph tells you how steep the line is.

Hertz: the unit of frequency.
1 Hz = 1 complete wave every second.

incident ray: this ray shows the light travelling towards the mirror.

independent variable: the variable in an investigation that you change.

infrasound: sound waves with a frequency too low for humans to hear.

light ray: straight line which shows the direction of light.

line of best fit: straight or curved line drawn through the middle of a set of points to show the pattern of data points.

longitudinal wave: in a longitudinal wave, the wave travels in the same direction as the vibrations that produced it.

magnetic field: the region around a magnetic material in which a magnetic force acts.

magnetic force: force that occurs when a magnet attracts another object or repels another magnet.

magnetic pole: point on a magnet where the force is strongest.

magnetism: property of some materials that gives rise to forces between these materials and magnets.

medium: the substance a sound wave travels through.

normal: this is a line drawn at 90° to the mirror at the point where rays hit the mirror.

opaque: material that prevents light travelling through.

permanent magnet: object made from a magnetic material that retains its magnetism for a very long time.

pitch: the pitch of a sound is how high or low it is.

plane mirror: plane means flat so a plane mirror is a flat mirror.

primary colours: red, blue and green. Mixing these colours of light together will make all other colours of light.

prism: transparent object that refracts light.

rarefaction: region where particles in a longitudinal wave are further apart.

real depth: how deep something really is.

reflected ray: this shows the light travelling away from the mirror after it has been reflected.

refraction: the bending of light when it enters a different medium.

repel: push further apart.

scattering: scattering happens when light is reflected from particles and uneven surfaces.

secondary colours: yellow, magenta and cyan.

shadow: dark area caused when light is blocked.

solar eclipse: when the Sun looks like it is completely or partly covered with a dark circle.

speed: how far something moves in a given time.

transparent: material that lets light through.

ultrasound: sound waves with a frequency too high for humans to hear.

vacuum: completely empty space.

vibration: when something moves back and forwards many times, we say it vibrates.

wavelength: the length of one complete wave.

Index

Index

Acknowledgements

The publishers wish to thank the following for permission to reproduce photographs. Every effort has been made to trace copyright holders and to obtain their permission for the use of copyright materials. The publishers will gladly receive any information enabling them to rectify any error or omission at the first opportunity.

(t = top, c = centre, b = bottom, r = right, l = left)

p2: Cheryl Moulton / Alamy Stock Photo; p3t: Elly Godfroy/Alamy Stock Photo; p3b: Frantisek Staud/Shutterstock; p5: National Geographic Creative/Alamy Stock Photo; p7: Science Photo Library; p9: Le Do/Shutterstock; p11: CP DC Press/Shutterstock; p12t: Robyn Mackenzie/Shutterstock; p12b: Oleksandra Naumenko/Shutterstock; p14t: bitt24/Shutterstock; p14b: Ted Pink/Alamy Stock Photo; p15: margouillat photo/Shutterstock; p16: ifong/Shutterstock; p18r: Stephen Dorey ABIPP/Alamy Stock Photo; p18l: joshya/Shutterstock; p27: MARTYN F. CHILLMAID/SCIENCE PHOTO LIBRARY; p30: Left Handed Photography/Shutterstock; p32: PjrStudio/Alamy Stock Photo; p33t: michaeljung/Shutterstock; p33b: Coprid/Shutterstock; p36: I'm friday/Shutterstock; p50: Mediscan/Alamy Stock Photo; p55: Wim Van Egmond/SCIENCE PHOTO LIBRARY; p56: Digital Photo/Shutterstock; p62: EYE OF SCIENCE/SCIENCE PHOTO LIBRARY; p69: Zoonar GmbH/Alamy Stock Photo; p71: Paul Quayle/Alamy Stock Photo; p78: Kjersti Joergensen/Shutterstock; p79: Yusnizam Yusof/Shutterstock; p81: Paul Fievez/Stringer/Getty Images; p82: Nathan King/Alamy Stock Photo; p83: Nattanee.P/Shutterstock; p95: Palo-ok/Shutterstock; p100: © NASA; p101: imagedb.com/Shutterstock; p104: charistoone-travel/Alamy Stock Photo; p104: Andrew Lambert/SCIENCE PHOTO LIBRARY; p107: Sorbis/Shutterstock; p108: everything possible/Shutterstock; p109: Yavuz Sariyildiz/Shutterstock; p110: Yevhen Vitte/Shutterstock; p112l: studiomode/Alamy Stock Photo; p112c & 149: ANDREW LAMBERT PHOTOGRAPHY/SCIENCE PHOTO LIBRARY; p112r: ANDREW LAMBERT PHOTOGRAPHY/SCIENCE PHOTO LIBRARY; p114l: ZaZa Studio/Shutterstock; p114c: Y Photo Studio/Shutterstock; p114r: vvoe/Shutterstock; p114b: GIPhotoStock/SCIENCE PHOTO LIBRARY; p117: Stanislav71/Shutterstock; p121: islandboy_stocker/Shutterstock; p123t: Khirman Vladimar/Shutterstock; p123c: Byjeng/Shutterstock; p123b: Lukas Hudac/Shutterstock; p124: Charles D Winters/SCIENCE PHOTO LIBRARY; p125: Love Silhouette/Shutterstock; p126r: Andrey Khachatryan/Shutterstock; p127t: Andrey Armyagov/Shutterstock; p127br: Michele Burgess/Alamy Stock Photo; p127bl: MarcelClemens/Shutterstock; p129: simonidadj/Shutterstock; p130t: Mimadeo/Alamy Stock Photo; p130b: IkeHayden/Shutterstock; p132tl: Dmitry Kainovsky/Shutterstock; p132tr: Manuel Ribeiro/Alamy Stock Photo; p132bl: rtem/Shutterstock; p132br: Ivaylo Ivanov/Shutterstock; p133: Dr Ajay Kumar Singh/Shutterstock; p137: Dana.S/Shuttestock; p138tr: GIPhotoStock/SCIENCE PHOTO LIBRARY; p138l: GIPhotostock/SCIENCE PHOTO LIBRARY; p138br: MARTYN F. CHILLMAID/SCIENCE PHOTO LIBRARY; p140: ANDREW LAMBERT PHOTOGRAPHY/SCIENCE PHOTO LIBRARY; p141: GIPhotoStock/SCIENCE PHOTO LIBRARY; p142: keantian/Shutterstock; p143: GIPhotoStock/SCIENCE PHOTO LIBRARY; p146t: GIPhotoStock/SCIENCE PHOTO LIBRARY; p146b: Paolo Gallo/Shutterstock; p148: Patricia Hofmeester/Shutterstock; p150t: Asianet-Pakistan/Shutterstock; p150b: ANDREW LAMBERT PHOTOGRAPHY/SCIENCE PHOTO LIBRARY; p157: Michael Doolittle/Alamy Stock Photo; p160: PACIFIC PRESS/Alamy Stock Photo; p161t: zatvornik/Shutterstock; p161b: Shutterstock; p162: Alhovik/Shutterstock; p166: Richard Baker/Getty Images; p167l: Jim West/Alamy Stock Photo; p167r: Jim West/Alamy Stock Photo; p168: Jan van der Hoeven/Shutterstock; p169l: Sue C/Shutterstock; p169r: Artesia Wells/Shutterstock; p173: GIPhotoStock X/Alamy Stock Photo; p174: eightstock/Shutterstock; p175: Alexander Mazurkevich/Shutterstock; p181: eightstock/Shutterstock; p182: Michael Wheatley/Alamy Stock Photo; p184: furtseff/Shutterstock; p185t: CNRI/SCIENCE PHOTO LIBRARY; p185ba: Yakov Oskanov/Shutterstock; p185bb: Claudio Saba/Shutterstock; p185bc: Nail Bikbaev/

Acknowledgements

Shutterstock; p185bd: John de la Bastide/Shutterstock; p185be: Crispus International/Shutterstock; p187t: Klaus Vedfelt/Getty Images; p187b: wxin/Shutterstock; p189: Australian Safety Transport Bureau; p197: ssuaphotos/Shutterstock; p199l: Zack Frank/Shutterstock; p199cl: Garsya/Shutterstock; p199cr: cribe/Shutterstock; p199r: Min C. Chiu/Shutterstock; p200: vixits/Shutterstock; p202: Ardo Holts/Alamy Stock Photo; p204: tea maeklong/Shutterstock; p212: SIMON FRASER/Getty Images; p213: GraphicsRF/Shuttesrtock; p214: David Arky/Getty Images; p216t: View Stock/Alamy Stock Photo; p216b: charistoone-images / Alamy Stock Photo; p217: charistoone-images / Alamy Stock Photo; p219: sciencephotos/Alamy Stock Photo; p220: Zig Zag Mountain Art/Shuttterstock; p222: Petar An/Shuttterstock; p223: Hung Chung Chih/Shutterstock.